London Zoo

List of Vertebrated Animals

Living in the Gardens of the Zoological Society of London 1866

.

London Zoo

List of Vertebrated Animals
Living in the Gardens of the Zoological Society of London 1866

ISBN/EAN: 9783337069780

Printed in Europe, USA, Canada, Australia, Japan

Cover: Foto ©berggeist007 / pixelio.de

More available books at **www.hansebooks.com**

LIST

OF

VERTEBRATED ANIMALS

LIVING IN

THE GARDENS

OF THE

ZOOLOGICAL SOCIETY

OF LONDON.

(FOURTH EDITION.)

PRINTED FOR THE SOCIETY,
AND SOLD AT THEIR HOUSE IN HANOVER SQUARE.

LONDON:
MESSRS. LONGMANS, GREEN, READER, AND DYER,
PATERNOSTER ROW.

TAYLOR AND FRANCIS.　　　　　[RED LION COURT, FLEET STREET.

[Price One Shilling and Sixpence.]

LIST OF THE SOCIETY'S PUBLICATIONS.

PROCEEDINGS OF THE COMMITTEE OF SCIENCE AND CORRESPONDENCE OF THE ZOOLOGICAL SOCIETY OF LONDON.
8vo. 2 vols.

			To Fellows.	To the Public.
Part I.	1830–31	1 vol. 8vo................... Price	4s. 6d. ...	6s.
,, II.	1832.	,, ... ,,	4s. 6d. ...	6s.

PROCEEDINGS OF THE ZOOLOGICAL SOCIETY OF LONDON.
8vo. 15 vols. and Index. (First Series.)

			Price to Fellows.	Price to the Public.				Price to Fellows.	Price to Public.
Part	I. 1833.	1 vol. 8vo.	4s. 6d. ...	6s.	Part	IX. 1841.	1 vol. 8vo.	4s. 6d. ...	6s.
,,	II. 1834.	,,	4s. 6d. ...	6s.	,,	X. 1842.	,,	4s. 6d. ...	6s.
,,	III. 1835.	,,	4s. 6d. ...	6s.	,,	XI. 1843.	,,	4s. 6d. ...	6s.
,,	IV. 1836.	,,	4s. 6d. ...	6s.	,,	XII. 1844.	,,	4s. 6d. ...	6s.
,,	V. 1837.	,,	4s. 6d. ...	6s.	,,	XIII 1845.	,,	4s. 6d. ...	6s.
,,	VI. 1838.	,,	4s. 6d. ...	6s.	,,	XIV. 1846.	,,	4s. 6d. ...	6s.
,,	VII. 1839.	,,	4s. 6d. ...	6s.	,,	XV. 1847.	,,	4s. 6d. ...	6s.
,,	VIII. 1840.	(Out of print.)			Index 1830–47.		,,	4s. 6d. ...	6s.

PROCEEDINGS OF THE ZOOLOGICAL SOCIETY OF LONDON.
8vo. 13 vols. and Index. (Second Series.)

			Without Illustrations.		With Illustrations.	
			To Fellows.	To the Public.	To Fellows. £ s. d.	To the Public. £ s. d.
Part	XVI. 1848.	1 vol. 8vo.	4s. 6d. ...	6s. ... Price	1 1 0 ...	1 7 6
,,	XVII. 1849.	,,	4s. 6d. ...	6s. ... ,,	1 1 0 ...	1 7 6
,,	XVIII. 1850.	,,	4s. 6d. ...	6s. ... ,,	1 7 6 ...	1 18 0
,,	XIX. 1851.	,,	4s. 6d. ...	6s. ... ,,	0 16 0 ...	1 1 0
,,	XX. 1852.	,,	4s. 6d. ...	6s. ... ,,	0 16 0 ...	1 1 0
,,	XXI. 1853.	,,	4s. 6d. ...	6s. ... ,,	0 18 0 ...	1 4 0
,,	XXII. 1854.	,,	4s. 6d. ...	6s. ... ,,	1 0 0 ...	1 6 0
,,	XXIII. 1855.	,,	4s. 6d. ...	6s. ... ,,	1 7 6 ...	1 18 0
,,	XXIV. 1856.	,,	4s. 6d. ...	6s. ... ,,	1 1 0 ...	1 7 6
,,	XXV. 1857.	,,	4s. 6d. ...	6s. ... ,,	1 1 0 ...	1 7 6
,,	XXVI. 1858.	,,	4s. 6d. ...	6s. ... ,,	1 12 0 ...	2 2 0
,,	XXVII. 1859.	,,	4s. 6d. ...	6s. ... ,,	1 12 0 ...	2 2 0
,,	XXVIII. 1860.	,,	4s. 6d. ...	6s. ... ,,	1 12 0 ...	2 2 0
Index 1848–60.		,,	4s. 6d. ...	6s. ... ,,	0 4 6 ...	0 6 0

THE ILLUSTRATIONS OF THIS SERIES ARE ALSO SOLD SEPARATELY IN FIVE VOLUMES, AS FOLLOWS:—

				To Fellows. £ s. d.	To the Public. £ s. d.
Mammalia	1 vol., containing	83 Plates	... Price	2 8 0 ...	3 3 0
Aves	2 vols.,	173 ,,	... ,,	4 15 0 ...	6 6 0
Reptilia et Pisces ...	1 vol.,	43 ,,	... ,,	1 3 0 ...	1 10 0
Mollusca..............	1 vol.,	51 ,,	... ,,	1 3 0 ...	1 10 0
Annulosa et Radiata	1 vol.,	90 ,,	... ,,	2 8 0 ...	3 3 0

LIST

OF

VERTEBRATED ANIMALS

LIVING IN

THE GARDENS

OF THE

ZOOLOGICAL SOCIETY

OF LONDON.

1866.

PRINTED FOR THE SOCIETY,
AND SOLD AT THEIR HOUSE IN HANOVER SQUARE.

LONDON:
MESSRS. LONGMANS, GREEN, READER, AND DYER,
PATERNOSTER ROW.

PRINTED BY TAYLOR AND FRANCIS,
RED LION COURT, FLEET STREET.

PREFACE TO THE FIRST EDITION.

THIS List, which has been drawn up, under my superin-tendence, by Mr. Louis Fraser, contained, as originally pre-pared, only the Vertebrated animals living in the Gardens on the 31st of December last, on which day it has been the practice for some years to take an accurate census of the Society's stock. The Council having determined to print it, for the use of the Fellows and other persons who take an interest in the Collection, I have carefully revised the whole, and endeavoured to make it as complete as possible by adding the species received since the beginning of the year. A living collection being liable to perpetual change, it cannot, of course, be expected that a list of this sort can be absolutely correct at any given moment; but I believe that the errors and omissions, as it at present stands, are not very numerous.

It will be observed that the Society's Collection, which is supposed to contain the most extensive series of living ani-mals in existence, embraces about 1450 specimens, illustrating 188 species of Mammals, 409 of Birds, 62 of Reptiles, and 23 of Fishes—altogether 682 species of Vertebrates. There is, besides these, a large series of Invertebrated animals of dif-ferent classes, kept in the House devoted to Aquaria, which varies much in composition and amount. In the "Scientific Index" to the List of Animals in the Society's Gardens, published in 1844 (which is, I believe, the last classified list issued), the number of species exhibited is given as 335—

namely 134 Mammals, 197 Birds, 3 Reptiles, and 1 Fish. This shows that we have made considerable progress since that period.

I may add that I am solely responsible for the arrangement and scientific nomenclature of this list and for the correct determination of the species. The difficulty, I may almost say the impossibility, in many cases, of ascertaining the specific names of living animals with perfect accuracy, is too well known to render apologies necessary in several cases where I may have failed in effecting this object.

PHILIP LUTLEY SCLATER,

Secretary to the Zoological Society of London.

11 Hanover Square,
   June 2nd. 1862.

PREFACE TO THE SECOND EDITION.

THE first edition of this Catalogue having been found of much use for reference, both by those who resort to the Collection for scientific purposes, and by the Society's numerous correspondents in foreign parts, the Council have authorized the preparation by Mr. Fraser, under my superintendence, of the present new edition, which gives the names of such animals as were living in the Gardens on the 1st of January last, together with those of the species that have been received since that period up to the end of May last.

The present Catalogue includes the names of 821 species of Vertebrates—namely, 229 Mammals, 468 Birds, 84 Reptiles, and 40 Fishes—illustrated by 2080 specimens in all. This, it will be observed, is a considerable increase over the number given in the former Catalogue, in which only 682 species of Vertebrates were enumerated.

P. L. S.

11 Hanover Square,
August 1st, 1863.

PREFACE TO THE THIRD EDITION.

In the present edition of this List, which has been prepared, as in the case of the former editions, under my superintendence, by Mr. Louis Fraser, it has been thought the simplest plan to insert the names of all the species that have been exhibited in the Society's Collection during the past year.

This will partly account for the increase of the numbers over those given in the last edition of the catalogue—though not altogether, as the fact is that the Society's Collection of living animals is being augmented year by year. That this is the case will be easily seen by the subjoined table, in which the numbers of specimens of the three highest classes of Vertebrates in the Society's Gardens on the first of January during the past ten years are given:—

January 1st,	1855.	1856.	1857.	1858.	1859.	1860.	1861.	1862.	1863.	1864.
Quadrupeds..	387	394	443	379	285	364	467	450	485	567
Birds	768	770	802	775	881	819	931	843	1114	1063
Reptiles	104	118	156	137	156	137	192	121	149	100
Total	1259	1282	1401	1291	1322	1320	1590	1414	1748	1730

If the numbers were taken in the middle of the summer, instead of on January 1st of each year, they would be larger, as during the autumn the duplicates, consisting principally of animals bred in the Society's Gardens during the year, are disposed of as far as possible.

PHILIP LUTLEY SCLATER,
Secretary to the Zoological Society of London.

11 Hanover Square.
Jan. 1st, 1865.

PREFACE TO THE FOURTH EDITION.

THE fourth edition of this List has been prepared on exactly the same principles as the third edition.

The names of all the Vertebrated animals added to the Society's collection during the years 1865 and 1866 have been inserted in their proper places; and the list in its present state may be considered to contain a complete record of all the species of Vertebrates that have been exhibited alive in the Society's Gardens in the years 1863, 1864, 1865, and 1866.

The total number of these is as follows :—

Mammals	339
Birds	721
Reptiles	73
Batrachians	25
Fishes	54
Total	1212

P. L. S.

11 Hanover Square,
July 9th, 1867.

CONTENTS.

LIST OF ANIMALS.

Class MAMMALIA.

Order QUADRUMANA.

Family SIMIIDÆ.

Genus TROGLODYTES.

1. *Troglodytes niger*, Geoff.   Chimpanzee.
   Hab. West Africa.

   - *a.* Female.   Purchased, Nov. 15, 1864.
   - *b.* Male.   Purchased, Aug. 15, 1865.
   - *c.* Purchased, Feb. 10, 1866.
   - *d.* Male.   Deposited, June 13, 1866.
   - *e.* Female.   Deposited, Aug. 17, 1866.
   - *f.* Male.   Presented by J. Snowdon Henry, Esq., Sept. 28, 1866.

Genus SIMIA.

2. *Simia satyrus*, Linn.   Orang-outang.
   Hab. Borneo.

   - *a.* Female.   Purchased, May 7, 1864.
   - *b.* Male ; *c.* Female.   Purchased, Sept. 4, 1866.

Genus SEMNOPITHECUS.

3. *Semnopithecus entellus* (Linn.).   Entellus Monkey.
   Hab. India.

   - *a.* Presented by Miss M. L. Gordon, July 23, 1864.
   - *b.* Presented by Capt. H. T. Forbes, Roy. Art., June 26, 1865.
   - *c.* Purchased, March 28, 1866.
   - *d.* Purchased, May 1, 1866.

B

4. *Semnopithecus cephalopterus* (Zimm.). Purple-faced Monkey.
Hab. Ceylon.

a. Purchased, Nov. 14, 1864.

5. *Semnopithecus maurus* (Schreb.). Moor Monkey.
Hab. Java.

a. Received, July 30, 1864.
b. Received in exchange, Feb. 8, 1866.

Genus COLOBUS.

6. *Colobus ursinus*, Ogilby. Ursine Colobus.
Hab. West Africa.

a. Purchased, April 23, 1864.

Genus CERCOPITHECUS.

7. *Cercopithecus callitrichus*, Is. Geoff. Green Monkey.
Hab. West Africa.

a. Presented by C. H. White, Esq., Oct. 1, 1864.
b. Purchased, Nov. 28, 1864.
c. Presented by W. Cheetham, Esq., Nov. 14, 1865.
d. Presented by H. B. Yeulett, Esq., Nov. 21, 1865.
e. Presented by — Hall, Esq., Jan. 6, 1866.
f, g. Presented by J. D. Cameron, Esq., R.M.S.S. Sup., St. Thomas, W. I., Jan. 15, 1866. From St. Kitts, W. I.
h. Purchased, Jan. 15, 1866. From St. Kitts, W.I.
i. Presented by Lieut. R. B. Wilkinson, R.N., Jan. 18, 1866.
j. Presented by John Willmott, Esq., Sept. 30, 1866.

8. *Cercopithecus lalandii*, Is. Geoff. Vervet Monkey.
Hab. South Africa.

a. Presented by Lieut. Chester Doughty, Aug. 15, 1864.
b. Deposited, Nov. 3, 1864.
c. Presented by J. C. Thompson, Esq., Nov. 30, 1864.
d. Female. Received in exchange, Dec. 15, 1864.
e. Presented by A. F. Green, Esq., April 15, 1865.
f. Presented by Capt. Saxon, June 28, 1865.
g. Presented by Miss Shuter, Aug. 5, 1865.
h. Deposited, June 2, 1866.
i, j. Presented by — Brewer, Esq., July 27, 1866.
k. Deposited, Oct. 18, 1866.

l, m. Deposited, Oct. 25, 1866.
n. Presented by the Rev. W. H. Drewett, Nov. 3, 1866.

9. *Cercopithecus griseo-viridis*, Desm. Grivet Monkey.
   Hab. North-east Africa.

   a. Purchased, Feb. 27, 1864.
   b. Deposited, Jan. 9, 1866.
   c. Presented by His Royal Highness the Prince of Wales, F.Z.S., Feb. 24, 1865.
   d. Deposited, Oct. 18, 1865.

10. *Cercopithecus albogularis*, Sykes. Sykes's Monkey.
    Hab. East Africa.

    a. Received in exchange, Dec. 15, 1864.
    b. Purchased, Jan. 12, 1865.
    c. Purchased, Aug. 22, 1865.
    d Deposited, May 22, 1866.

11. *Cercopithecus cephus*, Erxl. Moustache Monkey.
    Hab. West Africa.

    a. Purchased, Oct. 21, 1862.
    b. Purchased, Aug. 30, 1865.
    c, d. Purchased, Aug. 11, 1866.

12. *Cercopithecus mona*, Erxl. Mona Monkey.
    Hab. West Africa.

    a. Purchased, April 5, 1862.
    b. Received in exchange, March 9, 1865.

13. *Cercopithecus diana* (Linn.). Diana Monkey.
    Hab. West Africa.

    a, b. Purchased, June 14, 1864.
    c. Purchased, Aug. 30, 1865.
    d. Purchased, May 18, 1866.
    e. Purchased, July 18, 1866.

14. *Cercopithecus petaurista* (Schreb.). Lesser White-nosed Monkey.
    Hab. West Africa.

    a. Deposited, Oct. 21, 1864.
    b. Deposited, June 13, 1866.
    c. Purchased, July 18, 1866.
    d. Purchased, Aug. 11, 1866.

B 2

15. *Cercopithecus nictitans* (Linn.).   Hocheur Monkey
    Hab. West Africa.

    a. Purchased, Jan. 17, 1865.

16. *Cercopithecus ruber* (Gm.).   Patas Monkey.
    Hab. Abyssinia.

    a. Presented by G. E. Stanley, Esq., Aug. 1, 1863.
    b–d. Purchased, Dec. 27, 1864.
    e. Purchased, Aug. 11, 1866.

17. *Cercopithecus cynosurus*, Desm.   Malbrouck Monkey.
    Hab. West Africa.

    a. Purchased, May 3, 1865.
    b. Purchased, May 13, 1865.
    c. Presented by W. W. Vine, Esq., R.N., May 24, 1866.

18. *Cercopithecus erythrogaster*, Gray.   Red-bellied Monkey.
    Hab. West Africa.

    a. Purchased, March 16, 1866.   Specimen described and figured,
        P.Z.S. 1866, p. 168, Pl. XVI.

19. *Cercopithecus talapoin*, Erxl.   Talapoin Monkey.
    Hab. West Africa.

    a. Presented by Lieut. W. B. Bridges, R.N., Nov. 28, 1866.
        From the Congo River.

Genus CERCOCEBUS.

20. *Cercocebus æthiops* (Kuhl).   Mangabey Monkey.
    Hab. West Africa.

    a, b. Received in exchange, Aug. 16, 1864.
    c. Presented by A. F. Ridgway, Esq., Dec. 3, 1864.
    d. Purchased, Dec. 13, 1866.

21. *Cercocebus fuliginosus*, Geoff.   Sooty Monkey.
    Hab. West Africa.

    a. Presented by — Addison, Esq., April 13, 1864.
    b, c. Purchased, Aug. 30, 1865.
    d. Deposited, Nov. 24, 1865.
    e. Deposited, July 23, 1866.

22. *Cercocebus lunulatus* (Kuhl).   Lunulated Monkey.
    Hab. West Africa.

    a. Purchased, July 31, 1862.

23. *Cercocebus albigena* (Gray).   Grey-cheeked Monkey.
    Hab. West Africa.

    a. Purchased, Sept. 12, 1865.
    b. Purchased, Aug. 22, 1866.

Genus MACACUS.

24. *Macacus radiatus* (Shaw).   Bonnet-Monkey.
    Hab. Continental India.

    a. Purchased, 1853.
    b. Purchased, Sept. 26, 1862.   From Burmah.
    c. Presented by G. W. Robinson, Esq., Oct. 8, 1863.
    d. Presented by W. Freeman, Esq., April 8, 1864.
    e. Received, April 20, 1864.
    f. Presented by J. Seaton, Esq., July 18, 1864.
    g. Presented by Miss Hall, Oct. 5, 1864.
    h. Deposited, Oct. 17, 1864.
    i. Deposited, Jan. 9, 1865.
    j. Presented by A. M. Furby, Esq., March 27, 1865.
    k. Presented by W. F. Baynes, Esq., March 27, 1865.
    l. Presented by the Masters E. & H. Harraden, June 17, 1865.
    m. Presented by H. H. Horn, Esq., June 21, 1865.
    n. Presented by Thomas Cope, Esq., Sept. 20, 1865.
    o. Presented by — Hawkins, Esq., Sept. 23, 1865.
    p. Deposited, Nov. 6, 1865.
    q. Presented by Mrs. A. Ansell, Nov. 22, 1865.
    r. Presented by H. J. de Cataret, Esq., Jan. 1, 1866.
    s. Presented by S. Yorke Martin, Esq., Feb. 3, 1866.
    t. Presented by Edward Marsden, Esq., Feb. 7, 1866.
    u. Presented by Miss Rosalie Harraden, March 24, 1866.
    v. Presented by J. L. Gray, Esq., June 1, 1866.
    w. Presented by J. M. Marr, Esq., Sept. 29, 1866.
    x. Presented by the Earl of Harlington, Nov. 1, 1866.
    y. Presented by Mrs. Pigot, Nov. 27, 1866.
    z. Presented by the Rev. W. J. Richardson, Dec. 8, 1866.

25. *Macacus pileatus* (Shaw).   Toque Monkey.
    Hab. Ceylon.

    a. Presented by Capt. Kelsall, R.E., July 31, 1862.
    b. Deposited, Aug. 24, 1863.

c. Presented by Miss Maria Laura Ronald, Dec. 13, 1865.
d. Presented by Mrs. John Williams, Oct. 19, 1866.

26. *Macacus cynomolgus* (Linn.).   Macaque Monkey.
Hab. India.

a. Purchased, Sept. 2, 1862.   From Siam.
b. Presented by Capt. Wellstead, Oct. 6, 1862.   From Borneo.
c. Presented by Mons. Bernard, July 1, 1863.
d. Presented by Mons. Möller, Sept. 16, 1863.
e. Presented by Capt. Loutit, Nov. 20, 1863.
f. Presented by H. J. P. Cotton, Esq., Dec. 21, 1863.
g. Presented by T. J. Hutton Wood, Esq., Aug. 18, 1864.
h. Presented by Mrs. R. B. Beddome, Sept. 5, 1864.
i. Presented by Messrs. Ellis & Coy, Sept. 9, 1864.
j. Deposited, Nov. 3, 1864.
k. Presented by Capt. Maitland, R.N., Dec. 15, 1864.
l. Presented by W. Poinard, Esq., Feb. 18, 1865.
m. Presented by Cecil T. King, Esq., March 30, 1865.
n. Presented by T. Hayes, Esq., April 11, 1865.
o. Presented by R. Acton Paidoe, Esq., May 10, 1865.
p. Presented by — Swinson, Esq., Aug. 22, 1865.
q. Presented by Henry Jubber, Esq., Sept. 21, 1865.
r. Presented by George Newman, Sept. 25, 1865.
s. Presented by Capt. Taylor, Oct. 13, 1865.
t. Presented by Charles Oferberg, Esq., Nov. 17, 1865.
u. Presented by Arthur Donnithorne, Esq., Nov. 20, 1865.
v. Presented by J. W. Pyne, Esq., Jan. 23, 1866.
w. Presented by D. M. Smith, Esq., R.N., H.M.S. 'Conqueror,'
     Feb. 23, 1866.
x. Presented by the Rev. F. Gerald Vesey, May 10, 1866.
y. Presented by Miss E. Hand, May 30, 1866.
z. Presented by Messrs. George & White, June 7, 1866.
aa. Presented by Frederick Fox, Esq., July 24, 1866.
bb. Presented by W. B. Tegetmeier, Esq., Aug. 3, 1866.
cc. Presented by John Hosking, Esq., Jun., Oct. 10, 1866.
dd. Purchased, Oct. 19, 1866.
ee. Presented by T. N. Price, Esq., Oct. 24, 1866.

27. *Macacus cyclopis*, Swinhoe.   Round-faced Monkey.
Hab. Island of Formosa.

a. Male ; b. Female. Presented by Robert Swinhoe, Esq., F.Z.S.,
     H.B.M.'s Vice-Consul, Formosa, Sept. 4, 1862.

28. *Macacus erythræus* (Schreb.).   Rhesus Monkey.
Hab. Continental India.

a. Born in the Menagerie, Sept. 29, 1860.
b. Presented by Lady Sartorius, Nov. 19, 1861.

c. Deposited, Oct. 12, 1863.
d. Born in the Menagerie, June 3, 1864.
e. Presented by H. Houlder, Esq., Aug. 29, 1864.
f. Presented by Mrs. Paton, Oct. 8, 1864.
g. Deposited, Oct. 23, 1864.
h. Deposited, Jan. 14, 1865.
i. Presented by H. Hindley, Esq., Aug. 5, 1865.
j. Presented by J. D. Lees, Esq., Sept. 6, 1865.
k. Presented by J. S. Jarvis, Esq., Sept. 7, 1865.
l. Presented by D. White, Esq., Nov. 28, 1865.

29. *Macacus nemestrinus* (Linn.).   Pig-tailed Monkey.
Hab. Java.

a. Received in exchange, Dec. 15, 1864.
b. Purchased, July 19, 1865.
c. Presented by S. Youlton, Esq., March 31, 1866.
d. Purchased, Dec. 17, 1866.

30. *Macacus speciosus*, F. Cuv. Japanese Monkey.
Hab. Japan.

a. Purchased, June 3, 1864.

31. *Macacus inornatus*, Gray.   Bornean Ape.
Hab. Borneo.

a. Purchased, Feb. 21, 1866.   Specimen described and figured,
   P. Z. S. 1866, p. 202, Pl. XIX.

32. *Macacus inuus* (Linn.).   Barbary Ape.
Hab. North Africa.

a, b. Deposited, Oct. 24, 1864.

33. *Macacus silenus* (Linn.).   Wanderoo Monkey.
Hab. Malabar Coast.

a. Presented by Capt. Pocklington, 18th Regt., Sept. 2, 1863.
b. Presented by — Hellendaal, Esq., Dec. 23, 1865.
c, d. Presented by Col. Denison, May 3, 1866.

Genus CYNOCEPHALUS.

34. *Cynocephalus anubis*, F. Cuv.   Anubis Baboon.
Hab. West Africa.

a. Male.   Purchased, Nov. 16, 1860.

b. Presented by the Crystal Palace Company, Sydenham, July 6, 1865.

35. *Cynocephalus porcarius* (Bodd.). Chacma Baboon.
   Hab. South Africa.

   a. Male. Purchased, April 22, 1861.
   b. Deposited, Dec. 1, 1864.
   c. Presented by A. G. Smith, Esq., May 8, 1865.
   d. Presented by Major Lenon, Dec. 22, 1866.

36. *Cynocephalus hamadryas* (Linn.). Arabian Baboon.
   Hab. Aden.

   a. Female. Presented by Gordon Sandiman, Esq., June 30, 1860.

37. *Cynocephalus leucophæus*, Desm. Drill.
   Hab. West Africa.

   a, b. Purchased, Jan. 17, 1865.

38. *Cynocephalus papio*, Desm. Guinea Baboon.
   Hab. West Africa.

   a. Deposited, Sept. 30, 1862.
   b. Purchased, Jan. 17, 1865.
   c. Deposited, Oct. 31, 1865.

39. *Cynocephalus babouin*, Desm. Yellow Baboon.
   Hab. Africa.

   a. Purchased, April 18, 1864.
   b. Received, March 8, 1865.

Family CEBIDÆ.

Genus ATELES.

40. *Ateles marginatus*, Geoff. Chuva Spider Monkey.
   Hab. Brazil.

   a. Purchased, Oct 24, 1864.
   b. Purchased, May 20, 1865.

41. *Ateles frontatus*, Gray. Black-fronted Spider Monkey.
   Hab. Nicaragua.

   a. Purchased, Sept. 30, 1864.

b. Purchased, Oct. 29, 1864.
c. Purchased, Dec. 9, 1864.

42. *Ateles belzebuth* (Briss.).   Marimonda Spider Monkey.
Hab. Guiana.

   a. Purchased, Sept. 27, 1864.
   b. Purchased, Jan. 3, 1866.
   c. Deposited, May 7, 1866.

43. *Ateles ater*, F. Cuv.   Black-faced Spider Monkey.
Hab. Brazil.

   a. Purchased, Dec. 2, 1863.
   b. Purchased, March 31, 1866.
   c. Purchased, June 11, 1866.
   d. Purchased, Aug. 13, 1866.
   e, f. Purchased, Oct. 15, 1866.

44. *Ateles paniscus* (Linn.).   Red-faced Spider Monkey.
Hab. Brazil.

   a. Purchased, May 20, 1865.
   b. Received, July 13, 1866.

45. *Ateles grisescens*, Gray.   Grizzled Spider Monkey.
Hab. South America.

   a. Purchased, Oct. 24, 1869.   Specimen described by Dr. Gray,
       P. Z. S. 1865, p. 732.

46. *Ateles cucullatus*, Gray.   Hooded Spider Monkey.
Hab. South America.

   a. Purchased, June 30, 1865.

Genus MYCETES.

47. *Mycetes ursinus* (Humb.).   Brown Howler.
Hab. New Granada.

   a, b. Purchased, June 12, 1864.
   c. Purchased, June 27, 1866.

Genus CEBUS.

48. *Cebus apella* (Briss.).   Brown Capuchin Monkey.
Hab. Guiana.

   a. Presented by William Lloyd Esq., Aug. 26, 1864.

b, c. Purchased, Jan. 17, 1865.
d. Presented by the Rev. Dan. Greatorex, April 3, 1865.
e, f. Purchased, May 3, 1865.
g–i. Purchased, May 18, 1866.

49. *Cebus capucinus,* Geoff. Weeper Capuchin Monkey.
Hab. Brazil.

a, b. Purchased, Jan. 17, 1865.
c. Purchased, July 14, 1865.
d. Presented by Miss Jones, Sept. 27, 1865.
e. Purchased, June 11, 1866.
f. Purchased, June 22, 1866.
g, h. Purchased, July 18, 1866.
i, j. Purchased, Aug. 28, 1866.
k. Deposited, Sept. 29, 1866.
l. Deposited, Oct. 17, 1866.

50. *Cebus albifrons,* Geoff. White-fronted Capuchin Monkey.
Hab. South America.

a. Purchased, Jan. 17, 1865.
b. Purchased, Sept. 13, 1865.
c. Deposited, Feb. 26, 1866.

51. *Cebus fatuellus,* Erxl. White-whiskered Capuchin Monkey.
Hab. Brazil.

a. Purchased, 1864.

52. *Cebus hypoleucus,* Geoff. White-throated Sapajou.
Hab. Central America.

a. Presented by A. Yates, Esq., Aug. 2, 1864.
b. Purchased, Oct. 17, 1864.
c. Purchased, March 7, 1865.
d. Purchased, June 22, 1866.
e. Presented by Dr. Edward B. Bigg, R.N., H.M.S. ' Devastation,' C.M.Z.S., Aug. 8, 1866.

Genus LAGOTHRIX.

53. *Lagothrix humbolti,* Geoffr.). Humboldt's Lagothrix.
Hab. Upper Amazon.

a, b. Purchased, Oct. 2, 1863. Specimens figured, P. Z. S. 1863, p. 374, Pl. XXXI.

Genus PITHECIA.

54. *Pithecia satanas* (Hoffm.).   Jacket-Monkey.
Hab. Brazil.

 a. Purchased, March 8, 1864.
 b. Purchased, Nov. 18, 1865.
 c. Purchased, Oct. 15, 1866.

55. *Pithecia leucocephala* (Audeb.).   White-headed Saki.
Hab. Brazil.

 a. Presented by W. H. Barton, Esq., R.M.S. 'Wye,' May 15,
 1866.

Genus NYCTIPITHECUS.

56. *Nyctipithecus felinus*, Spix.   Feline Douroucouli.
Hab. Brazil.

 a. Purchased, May 29, 1857.
 b. Presented by F. Le Breton, Esq., May 12, 1864.
 c. Purchased, June 19, 1865.

Genus CALLITHRIX.

57. *Callithrix sciureus* (Linn.).   Squirrel Monkey.
Hab. Brazil.

 a. Purchased, March 15, 1864.
 b. Presented by Lieut. C. S. Candall, R.N., Nov. 30, 1864.
 c. Purchased, Aug. 27, 1864.
 d. Purchased, June 3, 1865.
 e, f. Presented by the Prince de Joinville, June 5, 1865.
 g, h. Purchased, July 15, 1865.
 i. Purchased, Sept. 1, 1866.
 j. Deposited, Nov. 7, 1866.

Genus HAPALE.

58. *Hapale jacchus* (Linn.).   Marmoset Monkey.
Hab. Bahia.

 a–d. Purchased, June 5, 1865.
 e. Deposited, Oct. 29, 1865.
 f. Presented by Miss Warren, March 28, 1866.
 g, h. Presented by C. A. Saunders, Esq., Aug. 3, 1866.
 i, j. Presented by J. Snowdon Henry, Esq., Sept. 28, 1866.
 k, l. Presented by Henry G. Coleman, Esq., Oct. 5, 1866.

59. *Hapale penicillata* (Geoff.).   Black-eared Marmoset.
    Hab. Brazil.

    a. Deposited, Oct. 13, 1864.
    b, c. Purchased, April 7, 1865.
    d–h. Presented by Alexander Collie, Esq., Sept 5, 1865.

60. *Hapale ursula* (Geoff.).   Negro Tamarin.
    Hab. Brazil.

    a. Presented by Comm. C. A. J. Aysham, R.N., Aug. 4, 1864.
    b. Purchased, Sept. 1, 1866.

61. *Hapale midas* (Linn.).   Red-handed Tamarin.
    Hab. Surinam.

    a. Received, July 13, 1866.

62. *Hapale œdipus* (Linn.).   Pinche Monkey.
    Hab. New Granada.

    a. Purchased, June 15, 1864.
    b. Presented by T. A. de Mosquese, Esq., Aug. 2, 1864.
    c, d. Purchased, June 11, 1866.

63. *Hapale rosalia* (Linn.).   Silky Monkey.
    Hab. Brazil.

    a. Deposited, Aug. 15, 1866.

Family LEMURIDÆ.

Genus LEMUR.

64. *Lemur varius*, Geoff.   Ruffed Lemur.
    Hab. Madagascar.

    a. Presented by John Fleming, Esq., Jan. 19, 1864.
    b. Purchased, July 18, 1865.

65. *Lemur niger*, Geoff.   Black Lemur.
    Hab. Madagascar.

    a. Male. Purchased, Nov. 10, 1864.
    b. Female. Purchased, May 3, 1861. The original of *Lemur leucomystax*, Bartl., described and figured, P. Z. S. 1862, p. 347, Pl. XLI.
    c. Female. Purchased, Jan. 7, 1865.

66. *Lemur albifrons*, Geoff.   White-fronted Lemur.
    Hab. Madagascar.

    a. Purchased, March 12, 1863.

67. *Lemur xanthomystax*, Gray.   Yellow-cheeked Lemur.
    Hab. Madagascar.

    a. Purchased, April 24, 1863.

68. *Lemur nigrifrons*, Geoff.   Black-fronted Lemur.
    Hab. Madagascar.

    a. Purchased, May 29, 1857.
    b. Presented by F. Le Breton, Esq., May 12, 1864.
    c. Hybrid. Between *Lemur nigrifrons*, Geoff., and *Lemur xan-thomystax*, Gray.   Born in the Menagerie, April 28, 1865.

69. *Lemur catta*, Linn.   Ring-tailed Lemur.
    Hab. Madagascar.

    a. Presented by Lieut. R. R. Cock, Oct. 13, 1862.

Genus GALAGO.

70. *Galago allenii*, Waterh.   Allen's Galago.
    Hab. Fernando Po.

    a. Presented by William Henry Ashmall, Esq., C.M.Z.S., June 30, 1863.

71. *Galago crassicaudata* (Geoff.).   Grand Galago.
    Hab. East Africa.

    a. Presented by Dr. Waghorn, O.C.D.D., July 26, 1864.
    b. Purchased, June 22, 1865.
    c. Purchased, Feb. 12, 1866.

72. *Galago garnettii* (Ogilby).   Garnett's Galago.
    Hab. Port Natal.

    a. Purchased, Nov. 20, 1863.
    b. Presented by Francis Chalmers, Esq., Sept. 23, 1865.

73. *Galago maholi*, A. Smith.   Maholi Galago.
    Hab. South Africa.

    a, b. Deposited, May 26, 1866.

Genus NYCTICEBUS.

74. *Nycticebus tardigradus* (Linn.).   Slow Loris.
   Hab. South China.

   a. Presented by Dr. Thomas Coghlan, July 8. 1863.
   b. Presented by the Babu Rajendra Mullick, C.M.Z.S., Dec. 22, 1864.

Family CHIROMYIDÆ.

Genus CHIROMYS.

75. *Chiromys madagascariensis* (Gm.).   The Aye-aye.
   Hab. Madagascar.

   a. Female.   Presented by Edward Mellish, Esq., Aug. 12,1862.

Order CHIROPTERA.

Family PTEROPODIDÆ.

Genus PTEROPUS.

76. *Pteropus medius.* Temm.   Frugivorous Bat.
   Hab. India.

   a. Presented by Edward W. Bagshot, Esq., Oct. 1, 1863.

Order INSECTIVORA.

Family TALPIDÆ.

Genus TALPA.

77. *Talpa europæa*, Linn.   Common Mole.
   Hab. British Islands.

   a. Presented by G. N. L. Austin, Esq., March 15, 1864.

Order CARNIVORA.

Family CANIDÆ.

Genus CANIS.

78. *Canis lupus,* Linn.   Common Wolf.
Hab. Europe.

  a. Male. Born in the Menagerie, May 19, 1859.
  b. Female. Presented by Capt. Fitzgerald, F.Z.S., Aug. 4, 1862.
  c. Male. Presented by the Prince de Joinville, Dec. 10, 1863.

79. *Canis anthus,* F. Cuv.   Abyssinian Wolf.
Hab. Abyssinia.

  a. Male. Purchased, 1848.

80. *Canis mesomelas,* Schreb.   Black-backed Jackal.
Hab. South Africa.

  a. Male.   Presented by Edward Spencer, Esq., Jan. 3, 1851.
  b. Female.   Presented by Dr. Hay, Jan. 15, 1858.
  c, d. Presented by E. L. Layard, Esq., F.Z.S., Oct. 31, 1864.
  e, f. Presented by E. L. Layard, Esq., F.Z.S., March 23, 1865.
  g. Presented by M. M. de Pass, Esq., April 15, 1865.

81. *Canis niloticus,* Geoff.   Egyptian Fox.
Hab. North Africa.

  a. Male.   Presented by J. A. Laing, Esq., July 6, 1860.

82. *Canis azaræ,* Pr. Max.   Azara's Fox.
Hab. South America.

  a. Female.   Born in the Menagerie, May 17, 1858.
  b. Purchased, Feb. 9, 1864.
  c, d. Presented by W. K. Martin, Esq., June 17, 1864.
  e. Presented by Mrs. Laird Warren, Sept. 9, 1864.

83. *Canis lagopus,* Linn.   Arctic Fox.
Hab. Norway.

  a–f. Presented by Capt. Stewart, Oct. 7, 1863.
  g, h. Deposited, Oct. 15, 1866.

84. *Canis argentatus*, Desm.   Silver Fox.
Hab. North America.

   a. Male.; *b*. Female. Presented by William G. Smith, Esq.,
      Secretary to the Hudson's Bay Company, Oct. 27, 1854.
   c. Deposited, Nov. 20, 1865.

85. *Canis fulvus*, Desm.   Red Fox.
Hab. North America.

   a. Presented by William Reid, Esq., Aug. 17, 1864.

86. *Canis velox*, Say.   Kit Fox.
Hab. North America.

   a. Purchased, Sept 29, 1865.

87. *Canis cerdo*, Linn.   Fennec Fox.
Hab. Egypt.

   a. Male; *b*. Female. Purchased, Aug. 21, 1858.
   c, d. Purchased, Oct. 10, 1866.

88. *Canis dingo*, Blumenb.   Dingo.
Hab. Australia.

   a. Male; *b*. Female.   Presented by — Hume, Esq., Aug. 20,
      1862.
   c, d. Presented by Dr. Mueller, C.M.Z.S., Dec. 3, 1864.   From
      Melbourne.
   e–h. Born in the Menagerie, April 7, 1865.
   i. Presented by Capt. Williams, June 6, 1865.
   j–n. Presented by Dr. Mueller, C.M.Z.S., Sept. 6, 1865.
   o. Deposited, Dec. 3, 1865.

89. *Canis familiaris*, Linn.   Domestic Dog.
Hab.——?

   a. Male (Esquimaux var.).   Presented by C. A. Sothern, Esq.,
      F.Z.S., June 3, 1865.

90. *Canis aureus*, Linn.   Common Jackal.
Hab. India.

   a. Presented by — Brown, Esq., Aug. 5, 1863.
   b. Presented by J. Millar, Esq., Sept. 10, 1863.
   c, d. Born in the Menagerie, March 13, 1864.

91. *Canis vulpes*, Linn.   Common Fox.
   Hab. British Islands.

   a. Presented by John Billing, Esq., Aug. 7, 1866.

Genus OTOCYON.

92. *Otocyon lalandii* (Desm.).   Lalande's Long-eared Fox.
   Hab. South Africa.

   a. Male.   Presented by J. J. Barry, Esq., Oct. 23, 1863.
   b, c. Females.   Presented by Josiah Rivers, Esq., July 25, 1864.

Family HYÆNIDÆ.

Genus HYÆNA.

93. *Hyæna crocuta*, Erxl.   Spotted Hyæna.
   Hab. South Africa.

   a. Male.   Presented by Edmund Gabriel, Esq., H.B.M.'s Commissioner at Loando, Angola, Aug. 22, 1860.
   b. Received, June 13, 1865.
   c. Purchased, Oct. 20, 1865.

94. *Hyæna brunnea*, Thunb.   Brown Hyæna.
   Hab. South Africa.

   a. Female.   Purchased, 1853.

95. *Hyæna striata*, Zimm.   Striped Hyæna.
   Hab. India.

   a. Female.   Presented by Edward Murray Cookesley, Esq., 22nd Regiment, May 27, 1862.

Family VIVERRIDÆ.

Genus PARADOXURUS.

96. *Paradoxurus typus*, F. Cuv.   Common Paradoxure.
   Hab. India.

   a. Presented by E. Lowry, Esq., Sept. 3, 1860.
   b. Presented by W. Wiggins, Esq., Feb. 16, 1865.

c

97. *Paradoxurus pallasii,* Gray.   Pallas's Paradoxure.
    Hab. India.

   a. Presented by J. Duplex, Esq., Dec. 10, 1859.
   b. Presented by G. Wakeman, Esq., July 28, 1865.   From Java.
   c. Presented by Mrs. Spencer, Aug. 3, 1865.

98. *Paradoxurus tytleri,* Blyth.   Tytler's Paradoxure.
    Hab. Andaman Islands.

   a. Presented by A. Grote, Esq., C.M.Z.S., May 20, 1865.

99. *Paradoxurus aureus,* F. Cuv.   Golden Paradoxure.
    Hab. Ceylon.

   a. Purchased, Aug. 6, 1860.

Genus NANDINIA.

100. *Nandinia binotata* (Reinw.).   Two-spotted Paradoxure.
     Hab. West Africa.

   a. Purchased, March 8, 1861.
   b. Presented by George Tidcombe, Esq., R.N., Aug. 22, 1864.

Genus SURICATA.

101. *Suricata zenik* (Gm.).   Suricate.
     Hab. South Africa.

   a. Purchased, Feb. 8, 1866.
   b. Presented by the Lord Bishop of Graham's Town, May 13, 1866.

Genus HERPESTES.

102. *Herpestes griseus* (Geoff.).   Grey Ichneumon.
     Hab. India.

   a. Presented by Charles Clifton, Esq., F.Z.S., Oct. 1, 1861.
   b. Purchased, Oct. 3, 1859.
   c. Deposited, July 6, 1863.
   d. Purchased, Nov. 8, 1864.
   e. Presented by C. J. Mason, Esq., Aug. 9, 1865.
   f. g. Presented by John Da Costa, Esq., Sept 8, 1865.
   h. Presented by C. E. Darley, Esq., Dec. 14, 1865.
   i. Presented by H. Simkins, Esq., Jan. 23, 1866.
   j. Presented by W. Penfold, Esq., Oct. 3, 1866.

103. *Herpestes fasciatus,* Desm.   Banded Ichneumon.
Hab. West Africa.

    a. Purchased, April 11, 1864.
    b. Presented by Spencer Chapman, Esq., Nov. 7, 1864.   From
        South Africa.

104. *Herpestes auropunctatus,* Hodgs.   Spotted Ichneumon.
Hab. Bengal.

    a. Male ; *b.* Female.   Purchased, Aug. 7, 1860.
    c. Presented by Capt. Salvin, Nov. 30, 1864.

105. *Herpestes paludosus,* Cuv.   Marsh Ichneumon.
Hab. South Africa

    a. Purchased, Aug. 14, 1862.
    b. Purchased, May 27, 1865.

106. *Herpestes pulverulentus,* Wagner.   Dusty Ichneumon.
Hab. South Africa.

    a. Purchased, Aug. 30, 1862.

Genus GENETTA.

107. *Genetta vulgaris* (Linn.).   Common Genet.
Hab. South Europe.

    a. Female.   Purchased, 1861.
    b. Male. Hybrid between this species and male *Genetta tigrina.*
        Born in the Menagerie, Oct. 5, 1859.
    c. Presented by T. Aldersey, Esq., Oct. 4, 1864.   From Cape
        Coast.

108. *Genetta pallida,* Gray.   Pale Genet.
Hab. West Africa.

    a. Presented by J. J. Monteiro, Esq., Oct. 4, 1863.

109. *Genetta senegalensis,* Fischer.   Senegal Genet.
Hab. West Africa.

    a, b. Presented by William Vare, Esq., Oct. 20, 1864.
    c. Deposited, May 23, 1865.   From the Gambia.

Genus VIVERRICULA.

110. *Viverricula indica* (Geoff.).   Indian Civet Cat.
Hab. India.

    a. Purchased, May 22, 1861.

b. Purchased, April 15, 1862.
c. Presented by F. W. Robinson, Esq., 60th Rifles, Sept. 19, 1863. From the Mysore Jungle, Madras Presidency.

Genus VIVERRA.

111. *Viverra civetta* (Schreb.). African Civet Cat.
Hab. Africa.

a. Presented by the late King of Portugal, F.Z.S., Nov. 15, 1855.
b. Presented by Edmund Gabriel, Esq., H.B.M.'s Commissioner at Loando, Angola, Aug. 22, 1860.
c, d. Presented by John Fleming, Esq., May 9, 1866.

Genus ARCTICTIS.

112. *Arctictis binturong* (Raffl.). Binturong.
Hab. Malacca.

a. Male. Presented by Mrs. Samuel Rawson, Oct. 24, 1855.
b. Deposited, Oct. 18, 1865.

Family FELIDÆ.

Genus FELIS.

113. *Felis leo*, Linn. Lion.
Hab. Africa and South-western Asia.

a. Female. Presented by Mr. Alderman Finnis, March 27, 1856. From Babylonia.
b. Male. Purchased, Oct. 28, 1859. From South Africa.
c. Male; d. Female. Purchased, March 8, 1862.
e. Female. Deposited, Nov. 29, 1864.
f. Female. Presented by William Cubitt, Esq., May 11, 1865.

114. *Felis tigris*, Linn. Tiger.
Hab. Eastern Asia.

a. Male; b. Female. Presented by Major Marston, April 3, 1858.
c. Female. Presented by Michael Scott, Esq., April 22, 1862.
d. Female. Received in exchange, May 22, 1865.
e, Male; f. Female. Presented by Col. Daly, Aug. 4, 1865.
g. Presented by L. Ashburner, Esq., Oct. 21, 1865.

115. *Felis onça,* Linn. Jaguar.
  Hab. South America.

  a. Male. Presented by W. D. Christie, Esq., F.Z.S., March 16, 1858.

116. *Felis hernandesii* (Gray). Mexican Jaguar.
  Hab. Mazatlan.

  a. Female. Presented by Miss Mary Knight, Nov. 28, 1857. Specimen described and figured P. Z. S. 1857, p. 278, Pl. LVIII.
  b. Born in the Menagerie, Sept. 23, 1864.

117. *Felis leopardus,* Linn. Indian Leopard.
  Hab. Asia.

  a. Male. Presented by Her Majesty the Queen, Feb. 16, 1860.
  b. Female. Presented by the late King of Portugal, F.Z.S., Sept. 24, 1856.
  c. Deposited, Oct. 21, 1865.
  d, e. Born in the Menagerie, June 9, 1866.

118. *Felis varia* (Gray). African Leopard.
  Hab. Africa.

  a. Male. Presented by Sir John H. Drummond Hay, K.C.B., C.M.Z.S., H.B.M.'s Minister Resident at Tangiers, May 24, 1858. From Morocco.
  b. Female. Presented by the King of Portugal, F.Z.S., Nov. 15, 1862.
  c. Deposited, Sept. 29, 1863.
  d, Presented by Mrs. Campbell, June 10, 1865.
  e, f. Born in the Menagerie, Aug. 11, 1865.

119. *Felis jubata,* Schreb. Cheetah.
  Hab. Africa and South-west Asia.

  a. Male. Purchased, Nov. 19, 1858. From South Africa.
  b. Female. Presented by H.R.H. the Prince of Wales, K.G., F.Z.S., July 5, 1862. From Nazareth, Syria.
  c. Purchased, Dec. 22, 1866.

120. *Felis macrocelis,* Temm. Clouded Tiger.
  Hab. Assam.

  a. Male. Purchased, May 16, 1854.
  b. Male. Purchased, March 12, 1862.

121. *Felis concolor*, Linn.   Puma.
    Hab. America.

    a. Male; *b*. Female.   Presented by W. D. Christie, Esq.,
      F.Z.S., Jan, 14, 1857.

122. *Felis serval*, Schreb.   Serval.
    Hab. Africa.

    a. Purchased, Dec. 9, 1863.   From South Africa.
    b. Purchased, Nov. 12, 1864.   From West Africa.
    c. Presented by Lieut. J. Plumridge, 3rd W. I. Regt., April
      24, 1865.

123. *Felis pardalis*, Linn.   Ocelot.
    Hab. America.

    a. Presented by W. D. Christie, Esq., F.Z.S., March 16, 1858.
    b. Presented by E. J. Longton, Esq., Jan. 17, 1863.
    c. Presented by Sir W. Clay, Bart., F.Z.S., Dec. 15, 1863.
    d. Presented by Dr. A. A. Blandy, May 3, 1866.
    e. Presented by G. S. Günther, Esq., July 13, 1866.
    f. Purchased, Aug. 13, 1866.
    g, h. Purchased, Oct. 15, 1866.
    i. Presented by Hugh Wilson, Esq., Nov. 4, 1866.

124. *Felis catus*, Linn.   Wild Cat.
    Hab. Scotland.

    a. Presented by the Earl of Seafield, F.Z.S., May 7, 1864.
    b. Presented by Miss Johnstone Douglas, Nov. 24, 1864.

125. *Felis chaus*, Güld.   Egyptian Cat.
    Hab. North Africa.

    a. Purchased, Nov. 11, 1862.

126. *Felis javensis*, Horsf.   Javan Cat.
    Hab. Eastern Asia.

    a. Presented by Lieut. H. Hand, R.N., Jan. 16, 1864.
    b. Purchased, Feb. 10, 1866.   From Formosa.

127. *Felis caracal*, Schreb.   Caracal.
    Hab. India.

    a. Deposited, Oct. 21, 1865.

Family MUSTELIDÆ.

Genus MUSTELA.

128. *Mustela canadensis*, Schreb.   Fisher Marten.
Hab. North America.

> *a.* Male; *b.* Female. Presented by Capt. David Herd, H.B.C.S., C.M.Z.S., Oct. 6, 1860.

129. *Mustela martes*, Linn.   Pine Marten.
Hab. British Islands.

> *a.* Presented by Lord Baynes, Oct. 18, 1864.
> *b.* Presented by John Francis, Esq., April 1, 1865.
> *c.* Deposited, April 30, 1866.   From Kerry, Ireland.

Genus GRISONIA.

130. *Grisonia vittata* (Schreb.).   Grison.
Hab. South America.

> *a.* Male.   Purchased, Dec. 24, 1860.
> *b.* Female.   Presented by — Gayleard, Esq., April 6, 1863.
> *c.* Purchased, Aug. 23, 1866.

Genus GALERA.

131. *Galera barbara* (Linn.).   Tayra.
Hab. South America.

> *a.* Presented by Dr. Wucherer, C.M.Z.S., Jan. 4, 1864.   From Bahia.

Genus LUTRA.

132. *Lutra vulgaris* (Linn.).   Common Otter.
Hab. British Islands.

> *a.* Male; *b.* Female. Presented by the Marquis of Bath, F.Z.S., June 15, 1861.
> *c.* Presented by John Henry Gurney, Esq., F.Z.S., March 28, 1864.
> *d.* Presented by Lord Huntingfield, F.Z.S., May 17, 1865.
> *e.* Presented by William Burnley Hume, Esq., Nov. 9, 1865.
> *f.* Presented by the Hon. Rowland Hill, M.P., April 17, 1866.
> *g.* Presented by F. Ware, Esq., May 12, 1866.

Genus MEPHITIS.

133. *Mephitis chilensis*, Geoff. Chilian Skunk.
Hab. Chili.

a. Presented by Mr. Edmonds, April 7, 1864.

134. *Mephitis americana* (Shaw). Skunk.
Hab. Hudson's Bay.

a. Presented by Capt. D. Herd, H.B.C.S., C.M.Z.S., Nov. 20, 1865.

Genus MELES.

135. *Meles taxus* (Schreb.). Common Badger.
Hab. British Islands.

a. Male. Presented J. H. Thrupp, Esq., May 22, 1860.
b. Female. Presented by the Duke of Richmond, April 22, 1862.
c. Female. Presented by John Boswell, Esq., Dec. 9, 1863.
d. Presented by Lord Garvagh, F.Z.S., March 4, 1865.

136. *Meles ankuma*, Temm. Sand Badger.
Hab. Japan.

a, b. Purchased, June 1, 1865.

Genus MELLIVORA.

137. *Mellivora capensis* (Schreb.). Cape Ratel.
Hab. South Africa.

a. Male. Presented by Capt. Tower, R.N., July 11, 1857.
b. Female. Presented by E. Wemyss, Esq., Nov. 30, 1861.

138. *Mellivora indica* (Shaw). Indian Ratel.
Hab. India.

a, b. Males. Presented by Arthur Grote, Esq., C.M.Z.S., April 26, 1862.

139. *Mellivora leuconota*, Sclater. West African Ratel.
Hab. West Africa.

a. Female. Received in exchange, Aug. 3, 1866. Specimen described and figured, P. Z. S. 1867, Pl. VIII.

Family URSIDÆ.

Genus CERCOLEPTES.

140. *Cercoleptes caudivolvulus* (Pall.). Kinkajou.
Hab. Demerara.

a. Presented by J. S. Smith, Esq., F.Z.S., July 8, 1862.
b. Female. Purchased, June 29, 1863.
c. Presented by Dr. Wucherer, C.M.Z.S., Jan. 4, 1864. From Bahia.
d. Purchased, June 27, 1866.
e. Presented by J. Lucie Smith, Esq., July 5, 1866.
f. Presented by Major Thompson, Dec. 1, 1866.

Genus PROCYON.

141. *Procyon lotor* (Linn.). Raccoon.
Hab. North America.

a. Male. Purchased, July 11, 1861.
b. Presented by W. H. Adams, Esq., May 6, 1863.
c, d. Presented by Egbert W. Cooper, Esq., 2nd W. I. Regt., Nov. 10, 1863.
e, f. Presented by H. O. Harris, Esq., Sept. 26, 1864.

142. *Procyon cancrivorus*, Geoff. Crab-eating Raccoon.
Hab. Tropical America.

a. Purchased, March 16, 1864.

Genus NASUA.

143. *Nasua nasica* (Linn.). Coati.
Hab. America.

a. Female. Purchased, Jan. 10, 1863.
b. Male. Purchased, June 29, 1863.
c. Presented by Capt. Pike, Jan. 17, 1865.
d. Presented by Capt. Henry W. Notley, June 16, 1865. From Para.
e. Deposited, Sept. 16, 1865.
f. Presented by Sir W. C. Trevelyan, Bart., F.Z.S., Sept. 20, 1865.
g. Presented by J. Baron, Esq., March 22, 1866.
h, i. Presented by Capt. F. H. Johnson, R.N., June 7, 1866. From Para.
j. Deposited, Aug. 31, 1866. From Surinam.

k. Female. Presented by — Spencer, Esq., Aug. 9, 1862. Red variety.

l. Purchased, Jan. 1, 1864. Red variety.

m. Presented by John Da Costa, Esq., Sept. 8, 1865. Red variety.

n. Deposited, Sept. 20, 1865. Black variety.

Genus Thalassarctos.

144. *Thalassarctos maritimus* (Linn.).   Polar Bear.
Hab. Greenland.

a. Female. Purchased, Sept. 28, 1846.
b. Male. Purchased, Oct. 1850.
c. Born in the Menagerie, Nov. 16, 1865.
d, e. Born in the Menagerie, Nov. 24, 1866.

Genus Ursus.

145. *Ursus arctos*, Linn.   Brown Bear.
Hab. North Europe.

a. Male. Purchased, June 1, 1852.
b. Female. Presented by Major Hon. W. C. W. Coke, M.P., F.Z.S.
c. Young, var. *collaris*. Deposited, April 9, 1862.
d. Presented by Viscount Ranelagh, May 10, 1864.
e, f. Presented by H. N. Alder, Esq., July 19, 1864. From Russia.
g. Var. *beringensis*, Midd. Presented by Capt. W. Beauchamp Seymour, R.N., C.B., Oct. 18, 1864. From Japan.
h. Received, July 5, 1865.
i. Presented by Charles Sidgreaves, Esq., July 18, 1865.
j. Received in exchange, Feb. 7, 1866.

146. *Ursus syriacus*, Ehrenb.   Syrian Bear.
Hab. Western Asia.

a. Female. Purchased, April 4, 1851.
b. Male. Presented by E. T. Rogers, Esq., H.B.M.'s Consul at Damascus, Sept. 21, 1864. From Syria.

147. *Ursus tibetanus*, F. Cuv.   Himalayan Bear.
Hab. North India.

a. Female. Presented by W. H. Russell, Esq., F.Z.S., Oct. 7, 1859.
b. Presented by H. O. Hebeler, Esq., 6th Regt. Foot, Sept. 19, 1864.

c. Purchased, Oct. 24, 1866. From Formosa, *Ursus formosanus*, Swinhoe.

148. *Ursus japonicus*, Schleg. Japanese Bear.
Hab. Japan.

a, b. Young Females. Purchased, April 1, 1863.

149. *Ursus americanus*, Pall. Black Bear.
Hab. North America.

a. Male. Presented by J. Wingfield Malcolm, Esq., Oct. 5, 1860.
b. Female. Presented by Capt. David Herd, H.B.C.S., C.M.Z.S., Oct. 11, 1861.
c. Presented by Capt. D'Arcy, R.N., Nov. 22, 1864.
d. Presented by — Keye, Esq., Jan. 15, 1866.
e. Deposited, March 1, 1866.
f. Presented by Capt. David Herd, H.B.C.S., C.M.Z.S., Nov. 8, 1866. From the Hudson's Bay Territory.

150. *Ursus malayanus*, Raffl. Malayan Bear.
Hab. Malacca.

a, b. Purchased, March 12, 1863.

Genus MELURSUS.

151. *Melursus labiatus* (Blainv.). Sloth Bear.
Hab. India.

a. Female. Presented by Lieut. James Howe Mardon, 66th Regt., Nov. 19, 1863.
b. Purchased, Oct. 21, 1865.

Order PINNIPEDIA.

Family PHOCIDÆ.

Genus PHOCA.

152. *Phoca vitulina*, Linn. Common Seal.
Hab. British Islands.

a. Purchased, Aug. 2, 1861.

153. *Phoca fœtida*, Müll. Ringed Seal.
Hab. North Sea.

a. Female. Purchased, April 17, 1862.

Family OTARIIDÆ.

Genus OTARIA.

154. *Otaria hookeri* (Gray).   Hooker's Sea-bear.
Hab. Patagonia.

a. Purchased, Jan. 25, 1866.

Order RODENTIA.

Family SCIURIDÆ.

Genus SCIURUS.

155. *Sciurus cinereus*, Linn.   Grey Squirrel.
Hab. North America.

a. Male; b. Female.   Presented by H. R. H. the Prince
Wales, K.G., F.Z.S., Nov. 6, 1861.

156. *Sciurus niger*, Linn.   Black Squirrel.
Hab. North America.

a, b. Purchased, March 30, 1864.

157. *Sciurus capistratus*, Bosc.   Capistrated Squirrel.
Hab. North America.

a. Purchased, Nov. 14, 1862.

158. *Sciurus vulgaris*, Linn.   Common Squirrel.
Hab. British Islands.

a–c. Deposited, Dec. 31, 1862.
d, e. Purchased, Dec. 12, 1863.
f. Presented by A. M. Furby, Esq., March 27, 1865.
g. Presented by G. V. Hill, Esq., Oct. 7, 1865.
h. Presented by E. Boyle, Esq., Jan. 2, 1866.

159. *Sciurus bicolor*, Sparrm.   Jelerang Squirrel.
Hab. India.

a, b. Deposited, Oct. 29, 1863.
c. Purchased, Aug. 30, 1865.

160. *Sciurus purpureus,* Zimm.   Malabar Squirrel.
   Hab. India.

   a. Presented by Lieut.-Col. Turnbull, July 10, 1862.
   b. Presented by Capt. Gideon, April 28, 1864.
   c. Presented by A. G. Dove, Esq., Oct. 24, 1864.
   d. Presented by Sir Arthur Cotton, Nov. 17, 1862.
   e. Presented by R. H. Peirce, Esq., Oct. 1, 1864.

161. *Sciurus plantani,* Ljungh.   Plantain-Squirrel.
   Hab. Java.

   a. Purchased, Sept. 17, 1862.

162. *Sciurus palmarum,* Linn.   Palm-Squirrel.
   Hab. India.

   a, b. Presented by Miss Emily Lamprell, July 7, 1862.
   c. Presented by Mrs. W. Sturrock, Sept. 8, 1862.
   d, e. Presented by N. A. Wells, Esq., Jan. 19, 1865.
   f, g. Presented by D. H. Macfarlane, Esq., April 25, 1866.

163. *Sciurus ludovicianus,* Custis.   Yellow-footed Squirrel.
   Hab. Texas.

   a. Purchased, March 17, 1865.
   b. Purchased, Jan. 1, 1866.

Genus XERUS.

164. *Xerus erythropus,* Geoff.   Red-footed Squirrel.
   Hab. West Africa.

   a, b. Purchased, June 22, 1866.

Genus SCIUROPTERUS.

165. *Sciuropterus volucellus* (Pall.).   American Flying Squirrel.
   Hab. North America.

   a. Male; b. Female.   Born in the Menagerie, Aug. 18, 1860.

Genus SPERMOPHILUS.

166. *Spermophilus citillus,* F. Cuv. et Geoff.   European Souslik.
   Hab. Europe.

   a–e. Purchased, Aug. 24, 1865.

Genus ARCTOMYS.

167. *Arctomys marmotta* (Linn.).   Alpine Marmot.
Hab. Europe.

  a. Presented by S. Stauffere, Esq., Feb. 10, 1859.
  b, c. Purchased, Nov. 10, 1866.

168. *Arctomys ludovicianus*, Ord.   Prairie Marmot.
Hab. North America.

  a. Female.   Presented by the Proprietors of 'The Field'
      Newspaper, Dec. 29, 1859.
  b. Male.   Presented by Capt. James Downie, July 23, 1863.
  c. Born in the Menagerie, Feb. 26, 1864.

169. *Arctomys empetra* (Schreb.).   Quebec Marmot.
Hab. North America.

  a. Presented by Lieut.-Col. Rhodes, June 11, 1862.
  b, c. Presented by the Hon. A. Gordon, April 30, 1864.
      From New Brunswick.
  d. Presented by E. Yeoman, Esq., June 28, 1865.

Family MYOXIDÆ.

Genus MYOXUS.

170. *Myoxus glis*, Schreb.   Fat Dormouse.
Hab. Europe.

  a. Purchased, Feb. 18, 1863.   From Poland.
  b. Purchased, April 17, 1864.   From Austria.
  c–e. Purchased, May 13, 1865.
  f–h. Purchased, Sept. 24, 1866.

171. *Myoxus nitela*, Schreb.   Garden-Dormouse.
Hab. Europe.

  a, b. Purchased, Sept. 24, 1866.

172. *Myoxus muscardinus* (Linn.).   Common Dormouse.
Hab. British Islands.

  a. Presented by N. L. Austin, Esq., Dec. 27, 1865.

Family CASTORIDÆ.

Genus CASTOR.

173. *Castor canadensis*, Kuhl. Canadian Beaver.
Hab. North America.

   a. Male. Presented by the Hudson's Bay Company, Oct. 26, 1861.
   b. Female. Presented by the Hudson's Bay Company, Oct. 7, 1862.
   c. Born in the Menagerie, July 14, 1864.
   d. Male. Received, Sept. 10, 1866.

Family MURIDÆ.

Genus MUS.

174. *Mus rattus*, Linn. Black Rat.
Hab. British Islands.

   a. Purchased, July 29, 1863.
   b. Presented by Charles Hampden Wigram, Esq., Nov. 25, 1864.
   c. Presented by W. M. Allfrey, Esq., July 30, 1866.
   d. Presented by Thomas Smith, Esq., Aug. 3, 1866.

Genus HAPALOTIS.

175. *Hapalotis mitchellii* (Ogilby). Mitchell's Hapalote.
Hab. Australia.

   a. Male; b. Female. Purchased, Sept. 24, 1862.
   c, d. Born in the Menagerie, Feb. 11, 1863.

Genus CRICETUS.

176. *Cricetus vulgaris*, Desm. Hamster.
Hab. Europe.

   a, b. Purchased, Nov. 1, 1864.
   c. Male; d. Female. Purchased, Feb. 2, 1865.
   e–h. Presented by Dr. G. Hartlaub, F.M.Z.S., Oct. 26, 1865.

Genus FIBER.

177. *Fiber zibethicus* (Linn.). Musquash.
Hab. North America.

   a. Presented by A. Downs, Esq., C.M.Z.S., Oct. 1, 1866. From Halifax, U.S.A.
   b. Presented by Capt. Herd, H.B.C.S., C.M.Z.S., Nov. 14, 1866.

Family DIPODIDÆ.

Genus DIPUS.

178. *Dipus ægyptius* (Hasselq.).   Jerboa.
   Hab. Egypt.

   a. Male.   Presented by the Hon. Mrs. Stuart, Sept. 9, 1861.
   b. Presented by B. Cochrane Willis, Esq., Feb. 5, 1862.

Family HYSTRICIDÆ.

Genus HYSTRIX.

179. *Hystrix africæ australis*, Peters.   South African Porcupine.
   Hab. Europe and Africa.

   a. Presented by H.R.H. the Duke of Edinburgh, Nov. 12, 1860.
      From South Africa.

180. *Hystrix leucura*, Sykes.   Indian Porcupine.
   Hab. India.

   a. Purchased, 1855.
   b, c. Presented by F. U. Rungell, Esq., July 8, 1862.
   d. Presented by A. Oswald Brodie, Esq., April 18, 1864.
      From Ceylon.
   e, f. Presented by Col. Thompson, Aug. 25, 1865.
   g, h. Born in the Menagerie, April 12, 1866.

181. *Hystrix malabarica*, Day.   Orange-quilled Porcupine.
   Hab. Malabar.

   a–c. Presented by Sir William Thomas Denison, K.C.B.,
      Dec. 22, 1864.   Specimens described and figured, P. Z. S.
      1865, p. 352, Pl. XVI.
   d. Born in the Menagerie, April 25, 1865.

182. *Hystrix javanica*, F. Cuv.   Javan Porcupine.
   Hab. Java.

   a. Received in exchange, May 2, 1860.
   b, c. Received in exchange, May 30, 1860.

183. *Hystrix grotii* (Gray).   Grote's Porcupine.
   Hab. Malacca.

   a. Presented by A. Grote, Esq., C.M.Z.S., April 18, 1866.
      Specimen described as *Acanthochœrus grotei*, by Dr. Gray,
      P. Z. S. 1866, p. 306, Pl. XXXI.

Genus ATHERURA.

184. *Atherura africana,* Gray.   Brush-tailed Porcupine.
   Hab. West Africa.

   a, b. Purchased, July 23, 1860.

Genus CERCOLABES.

185. *Cercolabes prehensilis* (Linn.).   Brazilian Tree-Porcupine.
   Hab. South America.

   a. Presented by R. W. Keate, Esq., F.Z.S., Aug. 9, 1862.
   b. Purchased, May 18, 1866.
   c. Received, Aug. 22, 1866.

186. *Cercolabes insidiosus* (Licht.).   Guiana Tree-Porcupine.
   Hab. Surinam.

   a. Received, July 13, 1866.

Genus CHINCHILLA.

187. *Chinchilla lanigera,* Benn.   Chinchilla.
   Hab. Chili.

   a. Male.   Born in the Menagerie, May 10, 1859.
   b. Female.   Purchased, May 3, 1865.
   c. Born in the Menagerie, May 31, 1866.

Genus LAGOSTOMUS.

188. *Lagostomus trichodactylus,* Brookes.   Viscacha.
   Hab. Buenos Ayres.

   a. Male.   Presented by Messrs. Werner & Co., Jan. 6, 1864.
   b–e. Presented by Capt. Parish, R.N., May 2, 1865.
   f. Presented by W. Bramley Moore, Esq., April 19, 1866.

Genus AULACODUS.

189. *Aulacodus swindernianus,* Temm.   Ground-Rat.
   Hab. West Africa.

   a. Purchased, June 13, 1866.

D

Genus ECHIMYS.

190. *Echimys spinosus,* Desm.   Spiny Rat.
    Hab. Brazil.

    a, b. Received in exchange from the Jardin des Plantes, Paris,
    June 26, 1865.

Genus CAPROMYS.

191. *Capromys pilorides,* Say.   Fournier's Capromys.
    Hab. Cuba.

    a. Purchased, May 19, 1862.
    b. Purchased, July 20, 1865.

192. *Capromys prehensilis,* Poepp.   Prehensile-tailed Capro-
    mys.
    Hab. Cuba.

    a. Purchased, June 18, 1862.

193. *Capromys brachyura,* Hill.   Short-tailed Capromys.
    Hab. Jamaica.

    a, b. Purchased, Oct. 23, 1865.

Genus MYOPOTAMUS.

194. *Myopotamus coypus* (Mol.).   Coypu.
    Hab. South America.

    a. Presented by Capt. Hutchkiss, Oct. 22, 1863.

Genus DASYPROCTA.

195. *Dasyprocta leporina* (Linn.).   Acouchy.
    Hab. British Guiana.

    a, b. Presented by Alfred Ikin, Esq., Nov. 2, 1861.
    c, d. Presented by Sir W. W. Holmes, July 10, 1866.

196. *Dasyprocta aguti* (Linn.).   Golden Agouti.
    Hab. Guiana.

    a. Female.   Presented by J. S. Caldwell, Esq., May 30, 1857.
    b. Male.   Born in the Menagerie, May 24, 1860.

c. Born in the Menagerie, June 28, 1863.
d. Born in the Menagerie, May 19, 1865.
e. Born in the Menagerie, Oct. 3, 1865.
f. Born in the Menagerie, March 7, 1866.
g. Presented by T. C. K. Cleeve, Esq., Oct. 15, 1866.
h. Born in the Menagerie, Dec. 6, 1866.

197. *Dasyprocta cristata,* Desm.   Crested Agouti.
Hab. South America.

a. Presented by Capt. M. D. Stewart, Oct. 12, 1861.
b, c. Presented by William Burnley Hume, Esq., Nov. 9, 1865.

198. *Dasyprocta* —— ?
Hab. South America.

a. Purchased, Oct. 13, 1862.
b. Purchased, June 3, 1865.
c. Purchased, Aug. 14, 1866.

Genus CŒLOGENYS.

199. *Cœlogenys paca* (Linn.).   Spotted Cavy.
Hab. Trinidad.

a. Presented by P. N. Bernard, Esq., Oct. 1, 1862.
b. Presented by Dr. Huggins, C.M.Z.S., Sept. 28, 1863.
c. Presented by the Comte and Comtesse d'Eu, Feb. 13, 1865.
e, f. Presented by R. T. Hadow, Esq., Oct. 19, 1865.
g. Presented by William Burnley Hume, Esq., Nov. 9, 1865.
h. Deposited, Jan. 23, 1866.
i. Presented by J. M. Barton, Esq., July 13, 1866.
j. Purchased, July 13, 1866.

Genus DOLICHOTIS.

200. *Dolichotis patachonica* (Shaw).   Patagonian Cavy.
Hab. Patagonia.

a, b. Presented by John Duguid, Esq., July 28, 1864.
c. Presented by Joseph Olguin, Esq., Sept. 28, 1866.

Genus CAVIA.

201. *Cavia caprera,* Linn., var.   Restless Cavy, or Guinea-Pig.
Hab. Brazil.

a, b, c, &c. Domestic variety.   Born in the Menagerie.

Genus HYDROCHŒRUS.

202. *Hydrochœrus capybara,* Erxl.   Capybara.
    Hab. South America.

    a. Female.   Purchased, June 13, 1859.
    b. Male.   Purchased, Feb. 3, 1863.
    c. Presented by Dr. Huggins, C.M.Z.S., Sept. 13, 1865.

Family LEPORIDÆ.

Genus LEPUS.

203. *Lepus mediterraneus,* Wagner.   Sardinian Hare.
    Hab. Africa.

    a. Deposited, April 21, 1863.

204. *Lepus europæus,* Pallas.   Common Hare.
    Hab. British Islands.

    a, b. Purchased, April 27, 1865.
    c. Presented by George Selwyn Morris, Esq., March **5**, 1866.

205. *Lepus timidus,* Linn.   Varying Hare.
    Hab. Northern Europe.

    a. Purchased, May 16, 1866.   From Ireland.

206. *Lepus cuniculus,* Linn.   Common Rabbit.
    Hab. British Islands.

    a. Male; *b, c, d.* Females.   Himalayan or Black-nosed
        variety.   Purchased, 1859.
    e, f. Males.   Moscow variety.   Deposited, Sept. 30, 1860.
    g, h, i. Males; *j, k, l.* Females.   Silver-grey variety.   De-
        posited, 1861.
    m. Female.   Variety from Porto Santo.   Deposited, 1861.

Order PROBOSCIDEA.

Family ELEPHANTIDÆ.

Genus ELEPHAS.

207. *Elephas indicus*, Linn.   Indian Elephant.
Hab. South India.

    a. Female.   Purchased, May 1, 1851.
    b. Male.   Young.   Presented by R. R. Cocq, Esq., Sept. 12, 1863.

208.   *Elephas africanus*, Blum.   African Elephant.
Hab. Africa.

    a. Male.   Received in exchange from the Jardin des Plantes, Paris, June 26, 1865.
    b. Female.   Purchased, Sept. 9, 1865.

Order ARTIODACTYLA.

Suborder RUMINANTIA.

Family CAMELIDÆ.

Genus AUCHENIA.

209. *Auchenia huanaco* (Mol.).   Huanaco.
Hab. Bolivia.

    a. Male; *b.* Female.   Presented by W. D. Christie, Esq., F.Z.S., March 16, 1858.
    c. Presented by Comm. Pinkerton, S.S. 'Thames,' Aug. 30, 1866.

210. *Auchenia pacos* (Linn.).   Alpaca.
Hab. Peru.

    a. Female.   Received in exchange, May 19, 1849.
    b. Male.   Received in exchange, Sept. 13, 1860.

211. *Auchenia glama* (Linn.).   Llama.
Hab. Peru.

    a. Male.   Received in exchange, April 28, 1860.
    b. Female.   Purchased, Jan. 21, 1861.

Genus CAMELUS.

212. *Camelus dromedarius*, Linn.   Common Camel.
Hab. Egypt.

 a. Male.   Presented by His late Highness Ibrahim Pasha,
 June 29, 1849.

213. *Camelus bactrianus*, Linn.   Bactrian Camel.
Hab. Central Asia.

 a. Female.   Presented by the Corps of the Royal Engineers,
 Nov. 18, 1856.   Born in the Crimea, 1855.

Family CAMELOPARDALIDÆ.

Genus CAMELOPARDALIS.

214. *Camelopardalis giraffa*, Gm.   Giraffe.
Hab. Kordofan.

 a. Male.   Born in the Menagerie, April 23, 1846.
 b. Female.   Born in the Menagerie, April 25, 1853.
 c. Female.   Born in the Menagerie, May 7, 1855.
 d. Male.   Born in the Menagerie, May 8, 1863.
 e. Male.   Born in the Menagerie, Sept. 24, 1863.
 f. Male.   Born in the Menagerie, March 31, 1865.
 g. Male.   Born in the Menagerie, April 20, 1865.
 h. Male.   Born in the Menagerie, Sept. 14, 1866.

Family ANTILOCAPRIDÆ.

Genus ANTILOCAPRA.

215. *Antilocapra americana* (Ord).   Pronghorn Antelope.
Hab. North America.

 a. Male.   Purchased, Jan. 21, 1865.

Family BOVIDÆ.

Genus OVIS.

216. *Ovis cycloceros*, Hutton.   Punjab Wild Sheep.
Hab. North-west India.

 a. Male.   Presented by Brigadier-General Hearsey, F.Z.S.,
 Aug. 19, 1854.

b. Female. Purchased, June 3, 1863.
c, d. Males. Born in the Menagerie, July 14, 1862.

217. *Ovis musimon*, Schreb. Mouflon.
Hab. Sardinia.

a. Male. Born in the Menagerie, May 1859.
b, c. Deposited, Aug. 13, 1866.

218. *Ovis aries*, Linn. Domestic Sheep.

a, b. Presented by John Henry Gurney, Esq., F.Z.S., Oct. 24, 1865. From Lagos.
c, d. Presented by Her Majesty the Queen, April 17, 1866. From Africa.
e, f. Presented by Capt. Glover, R.N., June 18, 1866. From Africa.

219. *Ovis tragelaphus*, Desm. Aoudad.
Hab. North Africa.

a. Male. Presented by H. E. Sir John Gaspard Le Marchant, G.C.M.G., Governor of Malta, March 2, 1861.
b. Female. Received in exchange, July 10, 1861.
c, d. Born in the Menagerie, March 16, 1864.
e, f. Born in the Menagerie, April 19, 1865.
g, h. Females. Born in the Menagerie, March 31, 1866.

Genus CAPRA.

220. *Capra megaceros*, Hutton. Markhoor.
Hab. Punjab.

a. Male. Presented by Capt. Samuel Browne, 2nd Punjab Cavalry, July 21, 1856.
b, c. Males. Born in the Menagerie, April 24, 1864. Hybrids between this species and female *Capra beden*.
d. Male; *e*. Female. Presented by Major F. R. Pollock, Commissioner at Dera Ismail Khan, Aug. 25, 1866.

221. *Capra beden*, Forsk. Cretan Goat.
Hab. Crete.

a. Female. Presented by F. Guarracino, Esq., H.B.M.'s First Vice-Consul, Constantinople, Dec. 16, 1862.
b. Hybrid. Born in the Menagerie, April 9, 1865.

222. *Capra ibex*, Linn. Ibex.
Hab. Savoy.

a. Male. Hybrid. Presented by the King of Italy, Nov. 3, 1862.

b. Female.　Hybrid.　Presented by the King of Italy, Nov. 3, 1862.
c. Hybrid.　Born in the Menagerie, April 6, 1863.
d. Hybrid.　Born in the Menagerie, May 22, 1865.
e. Hybrid.　Born in the Menagerie, April 3, 1866.

223. *Capra hircus,* Linn.　Domestic Goat.

a–k. Presented by the Babu Rajendra Mullick, C.M.Z.S., March 17, 1864.　Cashmere-Shawl Goats.
l, m. Presented by — Heley, Esq., Nov. 1, 1864.　Common Goats.
n. Born in the Menagerie, Jan. 23, 1865.　Cashmere-Shawl Goat.
o. Deposited, June 6, 1865.　Common Goat.
p. Born in the Menagerie, June 17, 1865.　Cashmere-Shawl Goat.
q. Born in the Menagerie, Jan. 31, 1866.　Cashmere-Shawl Goat.
r. Born in the Menagerie, Feb. 9, 1866.　Cashmere-Shawl Goat.
s. Born in the Menagerie, May 7, 1866.　Cashmere-Shawl Goat.
t. Born in the Menagerie, July 31, 1866.　Cashmere-Shawl Goat.
u. Born in the Menagerie, Aug. 7, 1866.　Cashmere-Shawl Goat.

Genus RUPICAPRA.

224. *Rupicapra tragus,* Gray.　Alpine Chamois.
Hab. Savoy.

a. Male ; *b.* Female.　Presented by the King of Italy, Nov. 3, 1862.
c. Born in the Menagerie, Aug. 6, 1864.
d. Born in the Menagerie, May 22, 1865.

Genus CEPHALOPHUS.

225. *Cephalophus maxwellii* (H. Smith).　Philantomba Antelope.
Hab. Sierra Leone.

a. Female.　Purchased, July 23, 1861.
b. Presented by Capt. J. L. Perry, R.N., of H.M.S. ‘Griffon,’ July 22, 1865.
c. Presented by Comm. St. Clair, R.N., H.M.S. ‘Sparrow,’ Aug. 17, 1866.

226. *Cephalophus pygmæus* (Linn.). Blaubok.
Hab. South Africa.

    a. Presented by His Royal Highness the Duke of Edinburgh, May 24, 1866.

227. *Cephalophus rufilatus*, Gray. Coquetoon Antelope.
Hab. West Africa.

    a. Female. Purchased, Aug. 27, 1861.

228. *Cephalophus breviceps*, Gray. Short-headed Antelope.
Hab. West Africa.

    a. Purchased, Feb. 13, 1866. Specimen described and figured, P. Z. S. 1866, p. 202, Pl. XX.

Genus CALOTRAGUS.

229. *Calotragus melanotis* (Rüpp.). Grys-bok.
Hab. South Africa.

    a. Purchased, June 24, 1864.

Genus HELEOTRAGUS.

230. *Heleotragus isabellinus*, Afz. Isabelline Antelope.
Hab. South Africa.

    a. Purchased, March 2, 1864.
    b. Female. Purchased, Nov. 28, 1865.

Genus PELEA.

231. *Pelea capreola* (Licht.). Reh-bok.
Hab. South Africa.

    a. Male. Presented by Edmund R. Wodehouse, Esq., April 29, 1863.
    b. Female. Purchased, June 24, 1864.

Genus DAMALIS.

232. *Damalis albifrons* (Burch.). Bless-bok Antelope.
Hab. South Africa.

    a. Female. Presented by H. E. Sir George Grey, K.C.B., F.Z.S., Governor of New Zealand, May 26, 1861.
    b. Male. Purchased, June 18, 1862.
    c. Female. Purchased, May 7, 1864.
    d. Born in the Menagerie, Jan. 29, 1866.

Genus ADENOTA.

233. *Adenota lechée*, Gray.   Lechée Antelope.
Hab. South Africa.

  a. Female.   Presented by H. E. Sir George Grey, K.C.B.,
      F.Z.S., Governor of New Zealand, March 1, 1859.

Genus GAZELLA.

234. *Gazella dorcas* (Linn.).   Gazelle.
Hab. Egypt.

  a. Male.   Presented by H. E. Sir John Gaspard Le Marchant,
      G.C.M.G., Governor of Malta, March 2, 1861.
  b. Presented by H. Straker, Esq., Feb. 2, 1864.
  c. Deposited, May 22, 1866.

235. *Gazella euchore* (Forst.).   Spring-bok.
Hab. South Africa.

  a. Female.   Purchased, Aug. 30, 1862.

236. *Gazella cuvierii*, Ogilby.   Cuvier's Gazelle.
Hab. Algeria.

  a. Male.   Purchased, Nov. 17, 1862.
  b. Presented by Capt. Alan Gardner, R.N., June 16, 1865.

237. *Gazella dama* (Pall.).   Dama Antelope.
Hab. Senegal.

  a. Male.   Purchased, Aug. 19, 1865.
  b. Female.   Purchased, Sept. 16, 1865.

238. *Gazella rufifrons*, Gray.   Korin.
Hab. Senegal.

  a. Purchased, Aug. 19, 1865.

Genus SAIGA.

239. *Saiga tatarica* (Pall.).   Saiga Antelope.
Hab. Siberia.

  a. Male.   Young.   Received in exchange, Nov. 21, 1864.
  b. Male.   Received, Jan. 21, 1865.
  c. Male ;   *d*. Female.   Purchased, Nov. 10, 1866.

Genus Hippotragus.

240. *Hippotragus niger*, Harris.   Sable Antelope.
Hab. South Africa.

   a. Male.   Purchased, Sept. 17, 1861.

Genus Boselaphus.

241. *Boselaphus caama* (Cuv.).   Hartebeest.
Hab. South Africa.

   a. Male.   Presented by H. E. Sir George Grey, K.C.B., F.Z.S., Governor of New Zealand, Nov. 1, 1861.

Genus Addax.

242. *Addax naso-maculatus* (Licht.).   Addax.
Hab. North Africa.

   a. Male.   Presented by H. E. Sir John Gaspard Le Marchant, G.C.M.G., Governor of Malta, March 2, 1861.
   b. Purchased, Sept. 30, 1864.

Genus Oryx.

243. *Oryx leucoryx* (Pall.).   Leucoryx.
Hab. North Africa.

   a. Female.   Born in the Menagerie, 1852.
   b. Male.   Born in the Menagerie, 1853.
   c. Female.   Born in the Menagerie, May 17, 1860.
   d. Female.   Born in the Menagerie, May 5, 1864.

Genus Tragelaphus.

244. *Tragelaphus scriptus* (Pall.).   Harnessed Antelope.
Hab. Gambia.

   a. Purchased, May 23, 1865.

Genus Oreas.

245. *Oreas canna* (Pall.).   Eland.
Hab. South Africa.

   a. Female.   Born in the Menagerie, Aug. 10, 1858.
   b. Female.   Born in the Menagerie, Dec. 20, 1861.
   c. Male.   Purchased, April 28, 1863.   From South Africa.

d. Male. Born in the Menagerie, June 9, 1863.
e. Female. Born in the Menagerie, June 14, 1863.
f. Male. Born in the Menagerie, Aug. 9, 1863.
g. Male. Born in the Menagerie, May 30, 1864.
h. Female. Born in the Menagerie, June 13, 1864.
i. Male. Born in the Menagerie, March 8, 1865.
j. Male. Born in the Menagerie, April 20, 1865.
k. Female. Born in the Menagerie, Feb. 22, 1866.
l. Male. Born in the Menagerie, Feb. 28, 1866.
m. Female. Presented by Lord Egerton of Tatton, Oct. 24, 1866.

Genus PORTAX.

246. *Portax picta* (Pall.). Nylghaie.
Hab. India.

a. Male. Born in the Menagerie, May 25, 1856.
b. Female. Purchased, Jan. 26, 1863.
c. Female. Deposited, Nov. 20, 1863.
d. Born in the Menagerie, Oct. 4, 1864.
e. Born in the Menagerie, Oct. 5, 1864.
f. Male. Purchased, Nov. 1, 1864.
g, h. Males. Born in the Menagerie, Sept. 15, 1866.
i, j. Born in the Menagerie, Oct. 7, 1866.

Genus CATOBLEPAS.

247. *Catoblepas gnu* (Gm.). White-tailed Gnu.
Hab. South Africa.

a. Male. Purchased, April 14, 1860.
b. Presented by William Cubitt, Esq., May 11, 1865.
c. Deposited, May 22, 1866.

248. *Catoblepas gorgon* (H. Smith). Brindled Gnu.
Hab. South Africa.

a. Female. Purchased, 1859.

Genus ANTILOPE.

249. *Antilope cervicapra*, Linn. Indian Antelope.
Hab. India.

a. Male. Purchased, Aug. 30, 1862.

Genus Bos.

250. *Bos grunniens*, Linn.   Yak.
   Hab. Tibet.

   a. Male ; b. Female.   Presented by C. M. Robinson, Esq., Feb. 21, 1861.
   c. Male.   Born in the Menagerie, June 20, 1863.
   d. Female.   Born in the Menagerie, July 9, 1864.
   e. Female.   Deposited, Feb. 21, 1865.
   f, g. Deposited, March 7, 1865.
   h. Deposited, July 17, 1866.

251. *Bos indicus*, Linn.   Zebu.
   Hab. Hindostan.

   a. Male ; b. Female.   Brahmin Cattle.   Presented by Her Majesty the Queen, May 17, 1862.
   c. Born in the Menagerie, March 7, 1863.
   d. Male.   Born in the Menagerie, March 2, 1865.

252. *Bos taurus*, Linn.   Domestic Ox.
   Hab. Italy.

   a. Male ; b. Female.   From Piedmont.   c. Male.   d. Female. From Tuscany.   Presented by the King of Italy, Nov. 3, 1862.
   e. Female.   Piedmont.   Born in the Menagerie, Sept. 14, 1863.
   f. Presented by G. Latimer, Esq., Sept. 2, 1865.   From the West Indies.

253. *Bos sondaicus*, Müll.   Sondaic Ox.
   Hab. Pegu.

   a. Male.   Presented by Col. A. P. Phayre, C.M.Z.S., May 4, 1863.

254. *Bos frontalis*, Lambert.   Gayal.
   Hab. India.

   a. Male.   Presented by the Babu Rajendra Mullick, C.M.Z.S., Dec. 29, 1865.

Genus Bison.

255. *Bison americanus* (Gm.).   American Bison.
   Hab. North America.

   a. Male.   Purchased, March 26, 1863.

b. Female. Purchased, April 30, 1864.
c. Male; *d.* Female. Purchased, Feb. 27, 1865.

Genus BUBALUS.

256. *Bubalus caffer,* Sparrm. Cape Buffalo.
Hab. South Africa.

a. Purchased, Nov. 6, 1865.

Family CERVIDÆ.

Genus CERVUS.

257. *Cervus canadensis,* Briss. Wapiti Deer.
Hab. North America.

a. Female. Born in the Menagerie, Sept. 17, 1851.
b. Female. Born in the Menagerie, July 25, 1853.
c. Male. Born in the Menagerie, Aug. 15, 1856.
d. Male. Var. *occidentalis,* H. Smith. Purchased, April 16, 1863. From Oregon.
e. Male. Born in the Menagerie, May 31, 1864.
f. Male. Born in the Menagerie, June 16, 1864.
g. Female. Born in the Menagerie, Aug. 17, 1865.
h. Female. Born in the Menagerie, Sept. 5, 1865.

258. *Cervus barbarus,* Benn. Barbary Deer.
Hab. North Africa.

a. Male. Presented by the Viscount Hill, F.Z.S., Feb. 29, 1860.
b. Female. Received in exchange, March 14, 1862.
c. Male; *d.* Female. Presented by Sir John Gaspard Le Marchant, G.C.M.G., Governor of Malta, Dec. 5, 1864.

259. *Cervus maral,* Gray. Persian Deer.
Hab. Persia.

a. Male; *b.* Female. Presented by the Earl of Ducie, F.Z.S., March 12, 1856.
c. Female. Born in the Menagerie, Aug. 23, 1858.
d. Male. Born in the Menagerie, Oct. 10, 1863.
e. Male. Born in the Menagerie, July 9, 1865.
f. Male. Born in the Menagerie, Aug. 29, 1865.
g. Male. Born in the Menagerie, Aug. 1, 1866.

260. *Cervus cashmirensis*, Gray.   Cashmirian Deer.
    Hab. Cashmir.

    a. Male.   Presented by Capt. M. H. S. Lloyd, H.M. 89th Regt.,
        Nov. 24, 1865.

261. *Cervus mantchuricus*, Swinhoe.   Mantchurian Deer.
    Hab. Newchang, China.

    a. Male.   Presented by R. Swinhoe, Esq., F.Z.S., H.B.M.'s
        Consul at Amoy, July 4, 1864.

262. *Cervus taivanus*, Blyth.   Formosan Deer.
    Hab. Island of Formosa.

    a. Male.   Presented by Robert Swinhoe, Esq., F.Z.S., H.B.M.'s
        Consul at Amoy, Dec. 9, 1861.
    b. Presented by J. J. Broadwood, Esq., June 23, 1865.
    c. Female.   Purchased, Feb. 10, 1866.

263. *Cervus sika*, Temm.   Japanese Deer.
    Hab. Japan.

    a. Male ; *b.* Female.   Presented by J. Wilks, Esq., July 21,
        1860.
    c. Female.   Purchased, June 5, 1861.
    d. Female.   Born in the Menagerie, Sept. 5, 1862.
    e. Female.   Born in the Menagerie, Aug. 2, 1863.
    f. Born in the Menagerie, June 8, 1865.
    g. Born in the Menagerie, June 26, 1865.
    h. Born in the Menagerie, June 3, 1866.
    i. Born in the Menagerie, July 8, 1866.

264. *Cervus duvaucelli*, Cuv.   Barasingha Deer.
    Hab. Himalaya.

    a. Female.   Purchased, Oct. 26, 1851.
    b. Female.   Born in the Menagerie, July 17, 1858.
    c. Male.   Born in the Menagerie, Aug. 26, 1861.
    d. Female.   Deposited, Dec. 5, 1863.
    e. Male.   Born in the Menagerie, June 13, 1864.

265. *Cervus aristotelis*, Cuv.   Sambur Deer.
    Hab. India.

    a. Female.   Born in the Menagerie, 1852.
    b. Female.   Born in the Menagerie, 1855.
    c. Male.   Born in the Menagerie, Aug. 1, 1862.

d. Male. Born in the Menagerie, April 30, 1863.
e. Born in the Menagerie, March 13, 1865.
f. Male. Born in the Menagerie, March 26, 1865.
g–i. Presented by Her Majesty the Queen, April 13, 1865.
j. Female. Received in exchange, July 5, 1865.
k. Female. Born in the Menagerie, Nov. 17, 1865.
l. Male. Born in the Menagerie, May 31, 1866.

266. *Cervus swinhoii*, Sclater. Swinhoe's Deer.
Hab. Island of Formosa.

a. Male. Presented by Robert Swinhoe, Esq., F.Z.S., H.B.M.'s Consul at Amoy, Sept. 4, 1862.

267. *Cervus rusa*, Müll. Rusa Deer.
Hab. Java.

a. Female. Received in exchange, March 1, 1851.
b. Male. Born in the Menagerie, Oct. 10, 1857.

268. *Cervus moluccensis*, Müll. Molucca Deer.
Hab. Molucca Islands.

a. Female. Purchased, June 18, 1862.
b. Male. Purchased, Sept. 16, 1863.
c. Male. Born in the Menagerie, Dec. 11, 1863.
d. Male. Born in the Menagerie, Nov. 8, 1864.
e. Born in the Menagerie, Sept. 30, 1865.

269. *Cervus timorensis*, Müll. Timor Deer.
Hab. Timor.

a. Male; *b.* Female. Presented by Capt. Lewis Brayley, Dec. 13, 1864.

270. *Cervus kuhlii*, Müller. Kuhl's Deer.
Hab. Bavian Islands.

a. Male; *b.* Female. Purchased, July 5, 1865.

271. *Cervus porcinus*, Zimm. Hog Deer.
Hab. India.

a. Female. Presented by H. E. Sir George Grey, K.C.B., F.Z.S., Governor of New Zealand, April 11, 1859.
b. Male. Born in the Menagerie, March 9, 1861.
c. Female. Purchased, Dec. 18, 1864.
d. Male. Born in the Menagerie, Feb. 23, 1866.

    e. Deposited, July 6, 1866.
    f. Born in the Menagerie, Nov. 10, 1866.

272. *Cervus axis,* Erxl.   Axis Deer.
    Hab. India.

    a. Female.   Presented by Richard C. Ansdell, Esq., F.Z.S.,
        May 5, 1859.
    b. Male.   Born in the Menagerie, Feb. 22, 1862.
    c. Female.   Presented by J. D. Mullins, Esq., Oct. 9, 1862.
    d. Male.   Born in the Menagerie, June 18, 1865.
    e. Male.   Presented by A. Houlder, Esq., Sept. 26, 1866.

273. *Cervus dama,* Linn.   Fallow Deer.
    Hab. British Islands.

    a. Male.   Deposited, 1863.
    b. Presented by Mrs. Randal Callander, Dec. 7, 1864.   From
        the Island of Rhodes.
    c. Female.   Purchased, Dec. 29, 1865.

274. *Cervus mexicanus,* H. Smith.   Mexican Deer.
    Hab. Mexico.

    a. Male.   Presented by A. Newton, Esq., F.Z.S., Sept. 3, 1857.
        From St. Croix, West Indies.
    b. Female.   Presented by Edward Sheldon, Esq., June 24,
        1863.   From Yucatan.
    c. Female.   Born in the Menagerie, Aug. 13, 1864.
    d. Male; *e.* Female.   Born in the Menagerie, Oct. 4, 1866.

275. *Cervus* ——?   North-American Deer.
    Hab. North America.

    a. Male.   Purchased, Aug. 27, 1861.
    b. Born in the Menagerie, July 2, 1865.
    c. Hybrid.   Between this species (*a*) and *Cervus mexicanus* (no.
        274, *b*).   Born in the Menagerie, Sept. 9, 1865.

276. *Cervus rufus,* F. Cuv.   Red Brocket.
    Hab. South America.

    a. Female.   Purchased, June 3, 1863.

277. *Cervus* ——?   Central-American Brocket.
    Hab. Honduras.

    a. Presented by Miss Williams, Nov. 19, 1864.

E

278. *Cervus pudu* (Mol.).   Pudu Deer.
    Hab. Chili.

    a. Male.   Presented by Charles Bath, Esq., Feb. 14, 18
       Specimen described, P. Z. S. 1866, p. 104.

Genus RANGIFER.

279. *Rangifer tarandus* (Linn.).   Reindeer.
    Hab. North Europe.

    a, b. Males.   Presented by H. H. Elder, Esq., June 1, 18

Family TRAGULIDÆ.

Genus TRAGULUS.

280. *Tragulus meminna* (Erxl.).   Indian Chevrotain.
    Hab. India.

    a. Presented by Capt. H. W. Hire, R.N., of H.M.S. 'Oront
       March 17, 1867.

281. *Tragulus javanicus* (Pall.).   Javan Chevrotain.
    Hab. Java.

    a. Presented by James A. Gardner, Esq., April 24, 1866.

Genus HYOMOSCHUS.

282. *Hyomoschus aquaticus* (Ogilby).   Water Chevrotain.
    Hab. West Africa.

    a. Purchased, Dec. 11, 1866.

Suborder NON-RUMINANTIA.

Family SUIDÆ.

Genus DICOTYLES.

283. *Dicotyles tajaçu* (Linn.).   Collared Peccary.
    Hab. South America.

    a. Male.   Presented by Andrew Arcedeckne, Esq., F.Z
       Sept. 28, 1860.
    b, c. Born in the Menagerie, Dec. 6, 1862.
    d. Born in the Menagerie, May 14, 1863.
    e. Deposited, June 18, 1863.

f. Born in the Menagerie, Oct. 16, 1863.
g. Presented by T. W. Sharpe, Esq., July 20, 1865.

284. *Dicotyles labiatus,* Cuv.   White-lipped Peccary.
    Hab. South America.

    a. Male.   Received in exchange, March 30, 1860.
    b. Hybrid.   Between this species and female *Dicotyles tajaçu*
        (Linn.).   Born in the Menagerie, Sept. 27, 1864.

Genus Sus.

285. *Sus scrofa,* Linn.   Wild Boar.
    Hab. Europe and North Africa.

    a. Male.   Var. *pliciceps,* Gray.   From China.   Deposited
        June 21, 1862.
    b. Female.   Presented by the Rev. B. Portal, July 3, 1862.
        From Italy.
    c. Presented by A. Christy, Esq., June 6, 1863.   From West
        Africa.
    d. Female.   Deposited, Jan. 20, 1865.   Europe.

286. *Sus* ——— ? ——— ?   Wild Pig.
    Hab. Dampier Straits, Eastern Archipelago.

    a. Presented by R. Swinhoe, Esq., F.Z.S., H.B.M.'s Consul
        at Amoy, July 2, 1864.

287. *Sus andamanensis,* Blyth.   Andaman Pig.
    Hab. Andaman Islands.

    a. Female.   Purchased, Feb. 28, 1863.
    b. Hybrid.   Between this species and *Sus,* no. 286.   Born in
        the Menagerie, Dec. 7, 1864.
    c–e. Presented by Capt. Frain, July 24, 1866.

288. *Sus taivanus,* Swinhoe.   Formosan Pig.
    Hab. Formosa.

    a–c. Received, Oct. 25, 1866.   Domestic brown variety.

289. *Sus leucomystax,* Temm.   White-whiskered Pig.
    Hab. Japan.

    a. Presented by Messrs. Glover & Co., May 30, 1865.

Genus BABIRUSA.

290. *Babirusa alfurus*, Less.   Babirusa.
Hab. Celebes.

a. Male. Received in exchange, Nov. 11, 1860.

Genus POTAMOCHŒRUS.

291. *Potamochœrus* penicillatus, Gray.   West-African River-Hog.
Hab. West Africa.

a. Female. Born in the Menagerie, June 4, 1858.
b. Male. Presented by M. du Chaillu, Nov. 18, 1864, as typical of his *P. albifrons.* From Gaboon.

Genus PHACOCHŒRUS.

292. *Phacochœrus æthiopicus* (Pall.).   Wart-Hog.
Hab. South Africa.

a, b. Presented by His Royal Highness the Duke of Edinburgh, May 6, 1866.

Family HIPPOPOTAMIDÆ.

Genus HIPPOPOTAMUS.

293. *Hippopotamus amphibius*, Linn.   Hippopotamus.
Hab. Upper Nile.

a. Male. Presented by the late Viceroy of Egypt, May 25, 1850.
b. Female. Presented by the late Viceroy of Egypt, July 22, 1854.

Order PERISSODACTYLA.

Family EQUIDÆ.

Genus EQUUS.

294. *Equus quagga*, Linn.   Quagga.
Hab. South Africa.

a. Female. Purchased, March 15, 1851.

b. Male. Presented by H. E. Sir George Grey, K.C.B., F.Z.S., Governor of New Zealand, Sept. 4, 1858.

295. *Equus burchellii*, Gray.   Burchell's Zebra.
Hab. South Africa.

a. Male. Purchased, 1850.
b. Female. Presented by H. E. Sir George Grey, K.C.B., F.Z.S., Governor of New Zealand, May 26, 1861.   Specimen figured, P. Z. S. 1865, pl. 22.
c. Male. Born in the Menagerie, July 6, 1865.

296. *Equus zebra*, Linn.   Common Zebra.
Hab. South Africa.

a. Presented by H. E. Sir Philip Edmund Wodehouse, K.C.B., Governor of the Cape Colony, July 27, 1864.

297. *Equus tæniopus*, Heugl.   African Wild Ass.
Hab. Abyssinia.

a. Purchased, Sept. 19, 1864.

298. *Equus onager*, Pall.   Asiatic Wild Ass, or Onager.
Hab. West Asia.

a. Female. Presented by the late W. Burkhardt Barker, Esq., Oct. 7, 1854.   From Syria.
b. Female. Presented by the Hon. C. A. Murray, C.B., F.Z.S., H.B.M.'s Envoy Ex. and Minister Plenipotentiary, March 11, 1859.   From Persia.
c. Female. Presented by Sir Thomas Erskine Perry, F.Z.S., May 5, 1849.   From Cutch.

299. *Equus hemionus*, Pall.   Tibetan Wild Ass, or Kiang.
Hab. Tibet.

a. Female. Presented by Major W. E. Hay, F.Z.S., Oct. 22, 1859.   Specimen figured P. Z. S. 1859, Mamm. pl. 73.

Family TAPIRIDÆ.

Genus TAPIRUS.

300. *Tapirus terrestris* (Linn.).   American Tapir.
Hab. South America.

a. Male. Received in exchange, Dec. 11, 1860.
b. Female. Presented by the late King of Portugal, F.Z.S., Feb. 14, 1857.

Family HYRACIDÆ.

Genus HYRAX.

301. *Hyrax capensis,* Schreb.   Hyrax.
Hab. South Africa.

  a. Presented by H.E. Sir George Grey, K.C.B., F.Z.S., Governor of New Zealand, Jan. 8, 1861.
  b-e. Purchased, Oct. 8, 1863.
  f, g. Born in the Menagerie, Nov. 25, 1863.
  h. Presented by Lieut.-Gen. Winyard, Sept. 17, 1863.
  i. Purchased, Sept. 28, 1866.

Family RHINOCEROTIDÆ.

Genus RHINOCEROS.

302. *Rhinoceros unicornis,* Linn.   Rhinoceros.
Hab. India.

  a. Female.   Purchased, July 17, 1850.
  b. Male ; *c.* Female.   Presented by A. Grote, Esq., C.M.Z.S., July 25, 1864.

Order EDENTATA.

Family BRADYPODIDÆ.

Genus BRADYPUS.

303. *Bradypus tridactylus,* Linn.   Three-toed Sloth.
Hab. British Guiana.

  a. Purchased, April 28, 1864.

Genus CHOLOPUS.

304. *Cholopus didactylus* (Linn.).   Two-toed Sloth.
Hab. Brazil.

  a, b. Purchased, May 10, 1865.

Family DASYPODIDÆ.

Genus DASYPUS.

305. *Dasypus encoubert,* Desm.   Weasel-headed Armadillo.
Hab. South America.

a, b. Purchased, June 30, 1860.
c. Deposited, Aug. 4, 1864.
d. Purchased, Aug. 30, 1864.
e, f. Purchased, May 11, 1865.

306. *Dasypus villosus,* Geoff.   Hairy Armadillo.
Hab. South America.

a. Presented by C. Bordas, Esq., Oct. 21, 1857.
b. Purchased, July 4, 1865.
c, d. Purchased, July 26, 1866.

307. *Dasypus minutus,* Desm.   Little Armadillo.
Hab. La Plata.

a, b. Purchased, Nov. 30, 1865.

308. *Dasypus peba,* Desm.   Peba.
Hab. South America.

a, b. Presented by Maximo Terrero, Esq., Oct. 14, 1864.

Genus TOLYPEUTES.

309. *Tolypeutes conurus,* Is. Geoff.   Three-banded Armadillo.
Hab. La Plata.

a. Received in exchange, March 4, 1865.

Order CETACEA.

Family DELPHINIDÆ.

Genus PHOCÆNA.

310. *Phocæna communis,* Less.   Porpoise.
Hab. British Waters.

a. Presented by Theodore Grant Crescy, Esq., Sept. 27, 1864.

b. Purchased, March 10, 1865. Specimen described by Dr. Gray under the name of *Phocæna tuberculifera,* P. Z. S. 1865, p. 318.
c. Purchased, Oct. 24, 1865.
d. Purchased, Dec. 1, 1865.
e–g. Purchased, April 16, 1866.

Order MARSUPIALIA.

Family DIDELPHYIDÆ.

Genus DIDELPHYS.

311. *Didelphys cancrivora,* Cuv.   Crab-eating Opossum.
Hab. Tropical America.

a. Presented by the Hon. F. North, Nov. 23, 1863.

312. *Didelphys virginiana,* Temm.   Virginian Opossum.
Hab. North America.

a. Presented by R. T. Hadow, Esq., June 21, 1866.

313. *Didelphys azaræ,* Temm.   Azara's Opossum.
Hab. South America.

a. Female; *b–d.* Young. Presented by William Reay, Esq. March 5, 1864.

Family DASYURIDÆ.

Genus DASYURUS.

314. *Dasyurus maugæi,* Geoff.   Maugé's Dasyure.
Hab. Australia.

a. Male.   Presented by H. E. Sir George Grey, K.C.B., F.Z.S. Governor of New Zealand, May 26, 1861.
b. Male.   Presented by F. J. C. Wildash, Esq., June 21, 1862
c. Presented by the Rev. E. Selwyn, Sept. 18, 1863.
d. Presented by W. S. C. Cooper, Esq., March 6, 1865.
e–h. Purchased, March 9, 1865.
i–l. Received, July 20, 1866.

315. *Dasyurus ursinus* (Harr.).   Ursine Dasyure.
Hab. Tasmania.

a. Male.   Presented by L. C. Stevenson, Esq., March 24, 1860

b. Female.　Presented by F. J. C. Wildash, Esq., Aug. 9, 1861.
c. Purchased, June 21, 1866.

Genus THYLACINUS.

316. *Thylacinus cynocephalus,* Harr.　Tasmanian Wolf.
Hab. Van Diemen's Land.

    a. Male.　Purchased, April 9, 1856.
    b. Male ; *c.* Female.　Presented by Ronald Gunn, Esq.,
        C.M.Z.S., May 2, 1863.

Family PHALANGISTIDÆ.

Genus PHALANGISTA.

317. *Phalangista vulpina* (Shaw).　Vulpine Phalanger.
Hab. Australia.

    a. Male.　Presented by James Selfe, Esq., Dec. 14, 1861.
    b. Presented by Frederick Moger, Esq., March 11, 1864.
    c. Presented by E. E. Ashley, Esq., Sept. 2, 1864.
    d–f. Presented by the Acclimatization Society of Victoria,
        Jan. 4, 1865.
    g. Born in the Menagerie, March 31, 1865.
    h. Presented by James R. Christie, Esq., Feb. 23, 1866.
    i. Deposited, April 19, 1866.
    j. Born in the Menagerie, July 17, 1866.
    k, l. Received, July 20, 1866.

318. *Phalangista fuliginosa,* Ogilby.　Sooty Phalanger.
Hab. Australia.

    a. Female.　Purchased, March 13, 1861.

319. *Phalangista canina,* Ogilby.　Short-eared Phalanger.
Hab. New South Wales.

    a. Deposited, Dec. 17, 1865.

Genus BELIDEUS.

320. *Belideus breviceps* (Waterh.).　Short-headed Phalanger.
Hab. Australia.

    a. Female.　Purchased, Oct. 15, 1861.
    b, c. Presented by W. T. Dayne, Esq., Oct. 19, 1862.
    d. Presented by George Macleay, Esq., F.Z.S., May 19, 1862.

    e. Presented by the Acclimatization Society of Victoria, Jan. 4,
       1865.
    f. Born in the Menagerie, June 8, 1865.
    g. Born in the Menagerie, June 28, 1866.

321. *Belideus flaviventer* (Geoff.).   Yellow-bellied Phalanger.
    Hab. New South Wales.

    a. Male; *b.* Female.   Presented, July 5, 1865.

Family PHASCOLOMYIDÆ.

Genus PHASCOLOMYS.

322. *Phascolomys wombat*, Pér. et Les.   Common Wombat.
    Hab. Tasmania.

    a. Presented by the Acclimatization Society of Victoria,
       March 18, 1863.
    b. Deposited, April 19, 1866.

323. *Phascolomys platyrhinus*, Owen.   Platyrhine Wombat.
    Hab. Victoria, Australia.

    a. Presented by the Acclimatization Society of Victoria,
       March 18, 1863,   Black variety, described by Mr. Gould
       as *P. niger.*

324. *Phascolomys latifrons*, Owen.   Hairy-nosed Wombat.
    Hab. Victoria, Australia.

    a. Purchased, July 24, 1862.   Specimen figured by Mr. Gould
       as *P. lasiorhinus.*

Family MACROPODIDÆ.

Genus DENDROLAGUS.

325. *Dendrolagus inustus*, Müll.   Brown Tree-Kangaroo.
    Hab. New Guinea.

    a. Received in exchange, Nov. 4, 1865.

Genus MACROPUS.

326. *Macropus rufus* (Desm.).   Red Kangaroo.
    Hab. Australia.

    a. Male.   Purchased, June 10, 1860.
    b. Female.   Purchased, May 25, 1866.

327. *Macropus melanops,* Gould.   Black-faced Kangaroo.
Hab. South Australia.

> *a.* Male.   Presented by T. R. Fletcher, Esq., April 29, 1859.
> *b.* Female.   Purchased, April 8, 1863.
> *c.* Received, Feb. 20, 1864.

328. *Macropus giganteus,* Shaw.   Great Kangaroo.
Hab. New South Wales.

> *a.* Male.   Presented by T. M. Mackay, Esq., April 11, 1863.
> *b.* Male.   Presented by the Acclimatization Society of Melbourne, Nov. 4, 1863.
> *c.* Presented by Mrs. Alexander Watson, March 16, 1864.
> *d.* Presented by Charles Magniac, Esq., June 10, 1865.
> *e.* Male.   Presented by F. C. Capel, Esq., Feb. 5, 1866.
> *f.* Born in the Menagerie, May 3, 1866.

Genus PETROGALE.

329. *Petrogale xanthopus,* Gray.   Yellow-footed Rock-Kangaroo.
Hab. South Australia.

> *a–d.* Purchased, April 2, 1864.
> *e, f.* Born in the Menagerie, April 4, 1865.

330. *Petrogale penicillata,* Gray.   Brush-tailed Kangaroo.
Hab. New South Wales.

> *a, b.* Purchased, May 25, 1866.
> *c.* Presented by H. Chandler, Esq., Oct. 24, 1866.

Genus HALMATURUS.

331. *Halmaturus bennettii,* Waterh.   Bennett's Wallaby.
Hab. Tasmania.

> *a–d.* Deposited, May 16, 1863.
> *e.* Male.   Received in exchange.
> *f, g.* Born in the Menagerie, May 26, 1865.
> *h.* Deposited July 1, 1865.
> *i.* Purchased, June 21, 1866.
> *j.* Born in the Menagerie, June 25, 1866.

332. *Halmaturus ruficollis* (Desm.).　Rufous-necked Wallaby.
Hab. New South Wales.

a, b. Females.　Presented by Thomas Maynard, Esq., June 8,
1856.
c. Female.　Presented by James Selfe, Esq., Dec. 14, 1861.
d. Male.　Hybrid.　Between this species and *Halmaturus
bennettii*, Waterh.　Born in the Menagerie, 1861.
e, f. Hybrids as above.　Born in the Menagerie, 1863.

333. *Halmaturus derbianus*, Gray.　Derbian Wallaby.
Hab. Australia.

a. Male ; b. Female.　Purchased, March 10, 1862.
c. Born in the Menagerie, 1863.
d. Purchased, Aug. 24, 1864.
e. Born in the Menagerie, May 26, 1865.
f. Presented by Dr. Mueller, C.M.Z.S., July 26, 1866.

Genus ONYCHOGALEA.

334. *Onychogalea lunata*, Gould.　Lunulated Kangaroo.
Hab. Australia.

a. Purchased, March 9, 1865.

Genus BETTONGIA.

335. *Bettongia grayi* (Gould).　Gray's Jerboa Kangaroo.
Hab. Australia.

a, b. Purchased, April 6, 1864.
c, d. Purchased, Jan. 27, 1863.
e. Born in the Menagerie, July 29, 1864.
f. Born in the Menagerie, May 3, 1866.
g. Presented by Robert Daubeney, Esq., July 16, 1866.
h. Born in the Menagerie, July 17, 1866.
i. Purchased, July 21, 1866.

336. *Bettongia* —— ?
Hab. Australia.

a. Presented by Messrs. Mitchell and W. Sluce, May 1, 1864.

337. *Bettongia rufescens*, Gray.　Rufous Jerboa Kangaroo.
Hab. New South Wales.

a. Purchased, March 9, 1865.

Family PERAMELIDÆ.

Genus PERAMELES.

338. *Perameles lagotis*, Reid.   Rabbit-eared Perameles.
Hab. West Australia.

a. Purchased, April 6, 1864.
b, c. Purchased, June 23, 1864.
d–f. Purchased, Feb. 14, 1866.

Order MONOTREMATA.

Family ECHIDNIDÆ.

Genus ECHIDNA.

339. *Echidna hystrix*, Cuv.   Echidna.
Hab. New South Wales.

a. Male.   Purchased, April 18, 1863.
b. Female.   Presented by E. T. Smith, Esq., July 18, 1863.

Class AVES.

Order PASSERES.

Family TURDIDÆ.

Genus TURDUS.

1. *Turdus musicus,* Linn.   Song-Thrush.
   Hab. British Islands.

   a. Purchased, 1861.

2. *Turdus viscivorus,* Linn.   Missel-Thrush.
   Hab. British Islands.

   a. Presented by Mr. Travis, July 1856.
   b. Presented by William Russell, Esq., F.Z.S., Sept. 15, 1862.

3. *Turdus migratorius,* Linn.   American Thrush.
   Hab. North America.

   a. Male; *b.* Female.   Purchased, April 23, 1859.
   c. Presented by William Russell, Esq., F.Z.S., Sept. 15, 1862.
   d. Presented by W. H. Boase, Esq., Sept. 14, 1864.

4. *Turdus merula,* Linn.   Blackbird.
   Hab. British Islands.

   a. Purchased, 1861.
   b. Presented by William Russell, Esq., F.Z.S., Sept. 15, 1862.

5. *Turdus torquatus,* Linn.   Ring-Ouzel.
   Hab. British Islands.

   a. Purchased, Nov. 15, 1860.
   b. Purchased, July 4, 1862.
   c, d. Females.   Presented by — Day, Esq., Jan. 21, 1864.

Genus PETROCINCLA.

6. *Petrocincla saxatilis* (Linn.).   Rock-Thrush.
   Hab. Europe.

   a. Purchased, April 26, 1862.

7. *Petrocincla cyanea* (Linn.). Solitary Thrush.
Hab. Europe.

  a. Male. Presented by Sir W. H. Fielden, Aug. 6, 1863.
  b. Purchased, Feb. 22, 1865.

Genus MIMUS.

8. *Mimus polyglottus* (Linn.). Mocking-bird.
Hab. North America.

  a. Purchased, 1860.

Genus ERITHACUS.

9. *Erithacus rubecula* (Lath.). Robin.
Hab. British Islands.

  a. Presented by William Russell, Esq., F.Z.S., Sept. 15, 1862.
  b. Presented by Mr. Travis, Dec. 1862.

Genus SIALIA.

10. *Sialia wilsonii*, Sw. Common Bluebird.
Hab. North America.

  a. Purchased, Aug. 29, 1862.
  b, c. Purchased, May 19, 1863.
  d–f. Purchased, Dec. 13, 1866.

Genus PHILOMELA.

11. *Philomela luscinia* (Linn.). Nightingale.
Hab. Europe.

  a. Purchased, 1865.

Family MOTACILLIDÆ.

Genus ANTHUS.

12. *Anthus arboreus*, Bechst. Tree-Pipit, or Titlark.
Hab. British Islands.

  a. Presented by William Russell, Esq., F.Z.S., Sept. 15, 1862.

13. *Anthus richardi*, Vieill.   Richard's Pipit.
    Hab. British Islands.

    a. Purchased, Oct. 8, 1866.

Genus MOTACILLA.

14. *Motacilla yarrellii*, Gould.   Pied Wagtail.
    Hab. British Islands.

    a, b. Purchased, Oct. 21, 1860.
    c, d. Bred in the Gardens, 1863.
    e, f. Bred in the Gardens, July 19, 1864.
    g, h. Purchased, June 4, 1866.
    i, j. Purchased, Oct. 8, 1866.
    k. Purchased, Oct. 22, 1866.

15. *Motacilla boarula*, Penn.   Grey Wagtail.
    Hab. British Islands.

    a–c. Purchased, Oct. 10, 1866.

Genus GRALLINA.

16. *Grallina australis*, Gray.   Pied Grallina.
    Hab. Australia.

    a. Male ; *b.* Female.   Presented by Dr. Mueller, C.M.Z.S.,
        Jan. 27, 1863.

Family PARIDÆ.

Genus SITTA.

17. *Sitta cæsia*, Meyer.   Common Nuthatch.
    Hab. British Islands.

    a. Purchased, April 30, 1860.

Genus LIOTHRIX.

18. *Liothrix luteus* (Scop.).   Yellow-bellied Leiothrix.
    Hab. India.

    a–d. Purchased, Sept. 4, 1866.

Family CRATEROPODIDÆ.

Genus GARRULAX.

19. *Garrulax chinensis* (Scop.).   Chinese Jay-Thrush.
Hab. China.

    a. Male. Presented by Robert Swinhoe, Esq., F.Z.S., H.B.M.
      Consul at Amoy, July 23, 1865.

Family PYCNONOTIDÆ.

Genus PYCNONOTUS.

20. *Pycnonotus pygæus* (Hodgs.).   Black Bulbul.
Hab. Bengal.

    a, b. Purchased, April 6, 1864.

21. *Pycnonotus jocosus* (Linn.).   Red-eared Bulbul.
Hab. India.

    a–e. Purchased, June 8, 1865.

22. *Pycnonotus crocorrhous,* Strickl.   Yellow-vented Bulbul.
Hab. India.

    a. Purchased, Sept. 7, 1865.

Family ARTAMIDÆ.

Genus ARTAMUS.

23. *Artamus superciliosus,* Gould.   White-eyebrowed Wood-
Swallow.
Hab. New South Wales.

    a, b. Presented by the Acclimatization Society of Melbourne,
      July 21, 1866.

Family DICRURIDÆ.

Genus CHIBIA.

24. *Chibia hottentotta* (Linn.).   Indian Drongo.
Hab. India.

    a. Purchased, Jan. 29, 1866.

F

Family CŒREBIDÆ.

Genus CŒREBA.

25. *Cœreba cyanea* (Linn.).   Yellow-winged Blue Creeper.
Hab. South America.

    a, b. Purchased, Dec. 14, 1864.

Genus DACNIS.

26. *Dacnis cayana* (Linn.).
Hab. Trinidad.

    a. Purchased, Aug. 7, 1866.

Family MELIPHAGIDÆ.

Genus PROSTHEMADURA.

27. *Prosthemadura novæ-zealandiæ* (Gm.). Poë Honey-eat
Hab. New Zealand.

    a. Presented by James M'Qaude, Esq., Aug. 12, 1865.
    b. Purchased, July 14, 1866.

Family TANAGRIDÆ.

Genus TANAGRA.

28. *Tanagra cana*, Sw.   Blue Tanager.
Hab. South America.

    a. Purchased, Aug. 20, 1864.

29. *Tanagra cyanoptera* (Vieill.).   Blue-shouldered Tanag
Hab. Brazil.

    a, b. Purchased, Oct. 29, 1864.

30. *Tanagra ornata*, Sparrm.   Archbishop Tanager.
Hab. Brazil.

    Purchased, Aug. 30, 1866.

Genus CALLISTE.

31. *Calliste fastuosa* (Less.).   Superb Tanager.
Hab. Pernambuco.

    a. Purchased, May 3, 1865.

32. *Calliste tricolor* (Gm.).   Green-headed Tanager.
Hab. South-east Brazil.

    a. Purchased, Nov. 24, 1865.
    b. Purchased, June 4, 1866.
    c, d. Purchased, Aug. 7, 1866.

Genus EUPHONIA.

33. *Euphonia violacea* (Linn.).   Violet Tanager.
Hab. Brazil.

    a–f. Purchased, Feb. 10, 1865.

34. *Euphonia nigricollis* (Vieill.).   Black-necked Tanager.
Hab. Brazil.

    a. Purchased, Aug. 7, 1866.

Genus PIPRIDEA.

35. *Pipridea melanota* (Vieill.).   Black-backed Tanager.
Hab. Brazil.

    a, b. Purchased, Aug. 7, 1866.

Genus RAMPHOCŒLUS.

36. *Ramphocœlus brasilius* (Linn.).   Brazilian Tanager.
Hab. Brazil.

    a. Male.   Purchased, July 14, 1863.

Genus DIUCOPIS.

37. *Diucopis fasciata* (Licht.).   Fasciated Tanager.
Hab. Para.

    a. Purchased, March 8, 1864.

Family FRINGILLIDÆ.

Subfamily ESTRELDINÆ.

Genus ESTRELDA.

38. *Estrelda ruficauda*, Gould.   Red-tailed Finch.
Hab. New South Wales.

a. Purchased, April 28, 1861.
b–d. Purchased, April 16, 1864.

39. *Estrelda phaëton*, Homb. et Jacq.   Crimson Finch.
Hab. Port Essington.

a, b. Males.   Presented by A. Denison, Esq., F.Z.S., June
1861.

40. *Estrelda bichenovii* (Jard. et Selb.).   Bicheno's Finch.
Hab. Queensland.

a–d. Presented by A. Denison, Esq., F.Z.S., June 5, 1861.

41. *Estrelda temporalis* (Lath.).   Australian Waxbill.
Hab. Australia.

a–d. Purchased, Nov. 4, 1861.
e–j. Presented by Dr. Mueller, C.M.Z.S., May 26, 1865.
k–z. Presented by Dr. Mueller, C.M.Z.S., June 15, 1865.

42. *Estrelda cinerea* (Vieill.).   Common Waxbill.
Hab. West Africa.

a, b. Purchased, May 8, 1860.
c, d. Deposited, Feb. 11, 1865.

43. *Estrelda amadava* (Linn.).   Amaduvade Finch.
Hab. India.

a, b. Deposited, Aug. 23, 1864.
c, d. Deposited, Feb. 11, 1865.

44. *Estrelda sanguinolenta*, Sw.   Zebra Waxbill.
Hab. Africa.

a. Deposited, Aug. 23, 1864.

45. *Estrelda phœnicotis,* Sw.   Crimson-eared Waxbill.
    Hab. West Africa.

> *a.* Purchased, Oct. 31, 1862.
> *b–d.* Deposited, Aug. 23, 1864.
> *e, f.* Deposited, Feb. 11, 1865.

46. *Estrelda melpoda* (Vieill.).   Orange-cheeked Waxbill.
    Hab. West Africa.

> *a, b.* Purchased, April 25, 1862.
> *c, d.* Deposited, Aug. 23, 1864.
> *e, f.* Deposited, Feb. 11, 1865.
> *g.* Deposited, Feb. 1, 1866.

47. *Estrelda rubriventris* (Vieill.).   Red-bellied Waxbill.
    Hab. West Africa.

> *a, b.* Purchased, May 29, 1862.
> *c.* Presented by Mrs. Gerrard Maynell, Dec. 8, 1864.

48. *Estrelda dufresnii* (Vieill.).   Dufresne's Waxbill.
    Hab. Natal.

> *a, b.* Purchased, Aug. 12, 1863.

49. *Estrelda* —— sp. ?
    Hab. St. Helena.

> *a, b.* Presented by Dr. R. Austen Allen, Oct. 14, 1865.

Genus QUELEA.

50. *Quelea occidentalis,* Hartl.   Occidental Finch.
    Hab. West Africa.

> *a, b.* Males ; *c, d.* Females.   Purchased, May 19, 1865.

Genus MUNIA.

51. *Munia undulata* (Lath.).   Nutmeg-Bird.
    Hab. India.

> *a–d.* Purchased May 29, 1862.

52. *Munia malacca* (Linn.).   Black-headed Finch.
    Hab. India.

> *a, b.* Purchased, May 29, 1862.

53. *Munia malabarica* (Linn.).　Indian Silver-bill.
Hab. India.

a–d. Purchased, July 4, 1862.

54. *Munia cantans* (Gm.).　African Silver-bill.
Hab. South Africa.

a. Purchased, 1860.
b. Deposited, Feb. 11, 1865.

55. *Munia acuticauda* (Hodgs.).　Sharp-tailed Finch.
Hab. India.

a–d. From Japan.　Purchased, Oct. 6, 1860.

56. *Munia striata* (Linn.).　Striated Finch.
Hab. India.

a. Purchased, Oct. 6, 1860.
b, c. Varieties from Japan.　Purchased, Oct. 6, 1860.

Genus PADDA.

57. *Padda oryzivora* (Linn.).　Java Sparrow.
Hab. Java.

a. Purchased, Oct. 6, 1860.
b–f. Presented by Charles Sidgreaves, Esq., April 10, 1864.

Genus POËPHILA.

58. *Poëphila cincta*, Gould.　Banded Grass-Finch.
Hab. Queensland.

a–d. Presented by A. Denison, Esq., F.Z.S., June 5, 1861.
e, f. Purchased, May 18, 1866.

Genus AMADINA.

59. *Amadina cucullata*, Swains.　Hooded Finch.
Hab. West Africa.

a–f. Deposited, Aug. 10, 1858.

60. *Amadina modesta,* Gould.   Modest Grass-Finch.
Hab. Australia.

    a. Purchased, April 25, 1862.

61. *Amadina castanotis,* Gould.   Chestnut-eared Finch.
Hab. Australia.

    a–f. Purchased, Oct. 18, 1862.

62. *Amadina fasciata* (Gm.).   Fasciated Finch.
Hab. West Africa.

    a. Male; *b.* Female.   Purchased, Jan. 4, 1864.

63. *Amadina lathami* (Vig. et Horsf.).   Spotted-sided Finch.
Hab. Australia.

    a–d. Purchased, April 16, 1864.
    e, f. Deposited, Aug. 23, 1864.
    g, h. Deposited, Feb. 11, 1865.
    i, j. Males; *k, l.* Females.   Purchased, June 16, 1865.
    m, n. Bred in the Gardens, Oct. 10, 1865.

Genus DONACOLA.

64. *Donacola castaneothorax,* Gould.   Chestnut-breasted
    Finch.
Hab. Queensland.

    a–d. Presented by A. Denison, Esq., F.Z.S., June 5, 1861.
    e, f. Purchased, April 28, 1861.

Subfamily VIDUINÆ.

Genus VIDUA.

65. *Vidua paradisea* (Linn.).   Whydah Bird.
Hab. West Africa.

    a. Male.   Purchased, Sept. 7, 1860.

66. *Vidua macroura* (Gm.).   Yellow-backed Whydah Bird.
Hab. West Africa.

    a. Male.   Purchased, Sept. 7, 1860.

Genus HYPHANTORNIS.

67. *Hyphantornis textor* (Gm.).   Rufous-necked Weaver-bird.
    Hab. West Africa.

    a–c. Purchased, Aug. 24, 1859.
    d. Presented by William Russell, Esq., F.Z.S., Sept. 15, 1862.
    e. Bred in the Gardens, Sept. 25, 1864.

68. *Hyphantornis castaneo-fuscus,* Less.   Chestnut-backed Weaver-bird.
    Hab. West Africa.

    a, b. Deposited, Aug. 19, 1854.

Genus PLOCEUS.

69. *Ploceus philippinensis* (Linn.).   Common Weaver-bird.
    Hab. India.

    a. Male; *b.* Female.   Purchased June 2, 1860.
    c, d. Deposited, Aug. 10, 1858.

70. *Ploceus melanogaster* (Lath.).   Black-bellied Weaver-Bird.
    Hab. West Africa.

    a–d. Purchased, July 18, 1862.

Genus EUPLECTES.

71. *Euplectes oryx* (Linn.).   Grenadier Weaver-bird.
    Hab. West Africa.

    a, b. Purchased, Sept. 17, 1862.
    c–f. Yellow variety.   Purchased, Sept. 2, 1865.

72. *Euplectes flammiceps,* Sw.   Crimson-crowned Weaver-bird.
    Hab. West Africa.

    a, b. Purchased, Feb. 20, 1864.

Genus TEXTOR.

73. *Textor alecto,* Temm.   Ox-bird.
    Hab. Senegal.

    a. Purchased, Aug. 19, 1865.

Subfamily EUSPIZINÆ.

Genus CYANOSPIZA.

74. *Cyanospiza cyanea* (Linn.).   Indigo-bird.
Hab. North America.

a–e. Purchased, Oct. 10, 1863.
f–h. Deposited, Aug. 23, 1864.
i, j. Deposited, Feb. 11, 1865.

75. *Cyanospiza ciris* (Linn.).   Nonpareil Finch.
Hab. North America.

a, b. Purchased, Aug. 12, 1863.
c. Deposited, Aug. 23, 1864.

Genus PHONIPARA.

76. *Phonipara olivacea* (Linn.).   Olive Finch.
Hab. Jamaica.

a. Male; b. Female.   Purchased, May 9, 1863.

77. *Phonipara canora* (Gm.).   Melodious Finch.
Hab. Cuba.

a, b. Males; c. Female.   Purchased, May 9, 1863.

78. *Phonipara bicolor* (Linn.).   Dusky Finch.
Hab. West Indies.

a. Purchased, Nov. 23, 1865.

Genus PAROARIA.

79. *Paroaria larvata* (Bodd.).   Red-headed Cardinal.
Hab. Brazil.

a, b. Presented by J. G. Leeming, Esq., Sept. 21, 1859.
c. Presented by Mrs. Croskey, Aug. 14, 1863.
d, e. Purchased, Dec. 26, 1863.
f. Purchased, Dec. 4, 1864.

Genus GUBERNATRIX.

80. *Gubernatrix cristatella* (Vieill.). Black-crested Cardinal.
Hab. South America.

a. Presented by Miss Williams Wynn, July 10, 1862.
b–d. Purchased, Dec. 13, 1866.

Genus TIARIS.

81. *Tiaris jacarina* (Linn.). Jacarini Finch.
Hab. South America.

a. Deposited, Aug. 10, 1858.

Subfamily FRINGILLINÆ.

Genus GUIRACA.

82. *Guiraca cærulea* (Linn.). Blue Grosbeak.
Hab. North America.

a, b. Purchased, July 31, 1862.

83. *Guiraca cyanea* (Linn.). Brazilian Blue Grosbeak.
Hab. Brazil.

a. Purchased, Nov. 23, 1865.

Genus CARDINALIS.

84. *Cardinalis virginianus* (Briss.). Cardinal Grosbeak.
Hab. North America.

a. Purchased, Aug. 29, 1862.
b, c. Purchased, April 20, 1865.
d–w. Deposited, July 13, 1865.
x. Purchased, April 4, 1866.

Genus ORYZOBORUS.

85. *Oryzoborus torridus* (Gm.). Tropical Seed-Finch.
Hab. South America.

a. Purchased, Aug. 8, 1860.

Genus MELOPYRRHA.

86. *Melopyrrha nigra* (Linn.).   Black Bullfinch.
Hab. Cuba.

    a. Male.   Purchased, May 9, 1863.

Genus SPERMOPHILA.

87. *Spermophila ophthalmica*, Sclater.   Spectacled Finch.
Hab. Ecuador.

    a. Female.   Purchased, Dec. 23, 1863.

88. *Spermophila cærulescens* (Vieill.).   Bluish Finch.
Hab. Brazil.

    a. Purchased, May 19, 1864.

89. *Spermophila albogularis* (Spix).   White-throated Finch.
Hab. Brazil.

    a. Purchased, May 19, 1864.

Genus PASSER.

90. *Passer montanus* (Linn.)   Tree-Sparrow.
Hab. British Islands.

    a. Purchased, Oct. 28, 1860.
    b. Presented by Mr. E. Bartlett, Dec. 31, 1862.
    c, d. Purchased, Dec. 19, 1864.

91. *Passer luteus* (Licht.).   Yellow Sparrow.
Hab. East Africa.

    a–d. Purchased, April 24, 1863.

92. *Passer simplex* (Licht.).   Grey-headed Sparrow.
Hab. West Africa.

    a–c. Purchased, Feb. 20, 1864.

Genus COCCOTHRAUSTES.

93. *Coccothraustes vulgaris* (Briss.).   Hawfinch.
Hab. British Islands.

    a. Purchased, Dec. 29, 1862.
    b. Purchased, Nov. 18, 1863.

c. Presented by C. Wolley, Esq., April 25, 1864.
d. Purchased, June 23, 1866.
e, f. Purchased, July 25, 1866.

Genus FRINGILLA.

94. *Fringilla chloris* (Linn.).   Greenfinch.
Hab. British Islands.

    a, b. Presented by Mr. E. Bartlett, Dec. 31, 1862.

95. *Fringilla aurantiiventris* (Cab.).   Algerian Greenfinch.
Hab. Algeria.

    a–c. Purchased, Aug. 1, 1864.

96. *Fringilla cœlebs,* Linn.   Chaffinch.
Hab. British Islands.

    a, b. Purchased, 1861.

97. *Fringilla spodiogenia,* Malh.   Algerian Chaffinch.
Hab. Algeria.

    a. Purchased, Aug. 1, 1864.

98. *Fringilla montifringilla,* Linn.   Mountain-Finch.
Hab. British Islands.

    a. Purchased, Dec. 3, 1859.
    b. Purchased, June 26, 1863.
    c. Male ; *d.* Female.   Purchased, Dec. 19, 1864.

99. *Fringilla serinus,* Linn.   Serin Finch.
Hab. Europe.

    a. Purchased, July 25, 1866.   Captured near London.

Genus ÆGIOTHUS.

100. *Ægiothus minor* (Ray).   Common Redpole.
Hab. British Islands.

    a. Presented by Mr. E. Bartlett, Dec. 31, 1862.

101. *Ægiothus canescens* (Gould). Mealy Redpole.
  Hab. British Islands.

  a. Purchased, Oct. 11, 1861.
  b. Presented by Mr. E. Bartlett, Dec. 31, 1862.
  c–h. Purchased, Oct. 30, 1866.

102. *Ægiothus montium* (Gm.). Mountain-Linnet, or Twite.
  Hab. British Islands.

  a, b. Purchased, Dec. 19, 1864.

Genus CARPODACUS.

103. *Carpodacus erythrinus* (Pall.). Ruddy Finch.
  Hab. Siberia.

  a–f. Purchased, Sept. 9, 1865.

Genus LINOTA.

104. *Linota cannabina* (Linn.). Linnet.
  Hab. British Islands.

  a–c. Purchased, Dec. 26, 1860.

Genus CORYTHUS.

105. *Corythus enucleator* (Linn.). Pine-Grosbeak.
  Hab. British Islands.

  a. Purchased, Aug. 16, 1864.
  b. Male ; *c.* Female. Presented by Robert Collett, Esq.,
    Sept. 12, 1866. From Christiana.

Genus LOXIA.

106. *Loxia curvirostra,* Linn. Crossbill.
  Hab. British Islands.

  a. Purchased, May 23, 1864.
  b. Purchased, June 23, 1866.
  c–h. Purchased, July 25, 1866.

Genus CARDUELIS.

107. *Carduelis elegans*, Steph.   Goldfinch.
Hab. British Islands.

a. Presented by Mr. E. Bartlett, Dec. 31, 1862.

Genus CHRYSOMITRIS.

108. *Chrysomitris tristis* (Linn.).   American Goldfinch.
Hab. North America.

a–i. Purchased, July 5, 1866.

Genus SYCALIS.

109. *Sycalis brasiliensis* (Gm.).   Saffron Finch.
Hab. Brazil.

a. Presented by William Russell, Esq., F.Z.S., Sept. 15, 1862.
b, c. Purchased, Dec. 4, 1864.

Genus CRITHAGRA.

110. *Crithagra butyracea* (Linn.).   St. Helena Seed-eater.
Hab. South Africa; St. Helena.

a. Male; b, c. Females.   Purchased, Sept. 24, 1862.
d. Deposited, Aug. 23, 1864.
e. Deposited, Feb. 11, 1865.
f–i. Purchased, April 12, 1866.

111. *Crithagra chrysopyga*, Sw.   Yellow-rumped Seed-eater.
Hab. West Africa.

a. Presented by F. Vanzeller, Esq., Nov. 15, 1862.

112. *Crithagra* —— sp.?
Hab. Africa.

a, b. Received in exchange, May 12, 1865.

Subfamily EMBERIZINÆ.

Genus PLECTROPHANES.

113. *Plectrophanes nivalis* (Linn.).   Snow-Bunting.
Hab. Europe and North Asia.

   a, b. Presented by A. Downs, Esq., C.M.Z.S., Aug 1, 1862.
      From Halifax.
   c–f. Purchased, Aug. 16, 1864.   From Siberia.

Genus CENTROPHANES.

114. *Centrophanes lapponica* (Linn.).   Lapland Bunting.
Hab. Europe.

   a. Purchased, Oct. 22, 1866.

Genus EMBERIZA.

115. *Emberiza hortulana*, Linn.   Ortolan Bunting.
Hab. British Islands.

   a–c. Purchased, Aug. 31, 1861.
   d. Presented by — Bicknell, Esq., April 26, 1865.

116. *Emberiza miliaria*, Linn.   Common Bunting.
Hab. British Islands.

   a–c. Purchased, 1861.

117. *Emberiza cirlus*, Linn.   Cirl Bunting.
Hab. British Islands.

   a–c. Purchased, Dec. 12, 1863.
   d. Purchased, June 6, 1866.

118. *Emberiza melanocephala*, Scop.   Black-headed Bunting,
Hab. South Europe.

   a. Female.   Presented by William Russell, Esq., F.Z.S.,
      Sept. 15, 1862.

119. *Emberiza schœniculus*, Linn.   Reed-Bunting.
Hab. British Islands.

   a. Purchased, Feb. 2, 1865.

Genus EUSPIZA.

120. *Euspiza luteola* (Sparrm.).   Red-headed Bunting.
Hab. India.

a–f. Presented by the Babu Rajendra Mullick, C.M.Z.S.,
Dec. 29, 1865.

Genus ZONOTRICHIA.

121. *Zonotrichia leucophrys*(Forst.). White-eyebrowed Finch.
Hab. Labrador.

a. Presented by Lewis Henry Spence, Esq., Dec 29, 1859.

Genus CALAMOPHILUS.

122. *Calamophilus biarmicus* (Linn.).   Bearded Reedling.
Hab. Europe.

a, b. Males.   Purchased, Oct. 1, 1861.
c. Female.   Purchased, Dec. 12, 1863.
d. Purchased, Aug. 16, 1865.
e–h. Purchased, Aug. 25, 1865.

Family ALAUDIDÆ.

Genus ALAUDA.

123. *Alauda arvensis,* Linn.   Skylark.
Hab. British Islands.

a. Female.   Presented by Mr. E. Bartlett, Dec. 31, 1862.

124. *Alauda arborea,* Linn.   Woodlark.
Hab. British Islands.

a. Purchased, Oct. 8, 1866.

125. *Alauda brachydactyla,* Temm.   Short-toed Lark.
Hab. Algeria.

a. Purchased, Aug. 1, 1864.

Genus MELANOCORYPHA.

126. *Melanocorypha mongolica* (Gm.). Chinese Lark.
Hab. China.

> *a.* Presented by Charles Rivington, Esq., June 1, 1866.

127. *Melanocorypha calandra,* Linn. Calandra Lark.
Hab. Europe.

> *a.* Deposited, Sept. 15, 1862.
> *b, c.* Deposited, May 11, 1865.

Family STURNIDÆ.

Subfamily ICTERINÆ.

Genus AGELÆUS.

128. *Agelæus phœniceus* (Linn.). Red-shouldered Starling.
Hab. North America.

> *a.* Purchased, April 13, 1860.
> *b.* Presented by A. Dixon, Esq., March 10, 1866.

Genus MOLOTHRUS.

129. *Molothrus badius* (Vieill.). Bay Cow-bird.
Hab. South Brazil.

> *a.* Purchased, July 13, 1859.

130. *Molothrus sericeus* (Licht.). Silky Cow-bird.
Hab. Trinidad.

> *a.* Male. Purchased, Sept. 29, 1862.
> *b.* Female. Purchased, Feb. 29, 1864.
> *c.* Purchased, Oct. 20, 1866.

Genus STURNELLA.

131. *Sturnella ludoviciana* (Linn.). Meadow Starling.
Hab. North America.

> *a.* Purchased, March 30, 1864.

G

132. *Sturnella defilippi* (Bonap.).   De Filippi's Starling.
     Hab. Rio de la Plata.

    a. Presented by Capt. A. Mellersh, R.N., May 2, 1865.

Genus CASSICUS.

133. *Cassicus persicus* (Linn.).   Yellow Hang-nest.
     Hab. Para.

    a, b. Purchased, March 8, 1864.

Genus CASSICULUS.

134. *Cassiculus melanicterus* (Bonap.).   Mexican Hang-ne
     Hab. Mexico.

    a. Presented by A. M. Booker, Esq., Aug. 18, 1865.

Genus ICTERUS.

135. *Icterus jamaicai* (Gm.).   Brazilian Hang-nest.
     Hab. Brazil.

    a. Male. Presented by Frederick Bernal, Esq., H.B.M
         Consul at Carthagena, New Granada, July 7, 1860.
    b. Purchased, May 3, 1865.
    c, d. Purchased, July 21, 1865.
    e, f. Purchased, Nov. 18, 1865.

Genus QUISCALUS.

136. *Quiscalus lugubris,* Sw.   Black Troupial.
     Hab. Trinidad.

    a–c. Purchased, Sept. 29, 1862.

Subfamily STURNINÆ.

Genus LAMPROCOLIUS.

137. *Lamprocolius chalybeus,* Ehrenb.   Green Glossy Starlin
     Hab. West Africa.

    a. Presented by Edward Cross, Esq., June 1, 1850.
    b–c. Purchased, April 15, 1865.

138. *Lamprocolius auratus* (Gm.).   Purple-headed Glossy
    Starling.
    Hab. West Africa.

    a. Purchased, March 14, 1856.
    b–d. Purchased, April 15, 1865.

139. *Lamprocolius rufiventris*, Rüpp.   Rufous-vented Glossy
    Starling.
    Hab. Western Africa.

    a, b. Purchased, Aug. 19, 1866.

Genus LAMPROTORNIS.

140. *Lamprotornis æneus* (Linn.).   Long-tailed Glossy
    Starling.
    Hab. West Africa.

    a–d. Purchased, April 15, 1865.

Genus STURNUS.

141. *Sturnus vulgaris*, Linn.   Common Starling.
    Hab. British Islands.

    a, b. Caught in the Gardens, 1861.

Genus PASTOR.

142. *Pastor roseus* (Linn.).   Rose-coloured Pastor.
    Hab. India.

    a–i. Presented by the Babu Rajendra Mullick, C.M.Z.S.,
    July 25, 1864.

Genus ACRIDOTHERES.

143. *Acridotheres cristatellus* (Linn.).   Chinese Mynah.
    Hab. China.

    a, b. Purchased, Aug. 24, 1859.
    c, d. Purchased, Sept. 9, 1865.

144. *Acridotheres ginginianus* (Lath.).　Indian Mynah.
Hab. Hindostan.

    a. Purchased, June 30, 1860.
    b. Presented by P. S. Laing, Esq., Oct. 11, 1862.
    c. Deposited, May 23, 1865.

145. *Acridotheres tristis* (Linn.).　Paradise Mynah.
Hab. India.

    a, b. Purchased, Sept. 9, 1865.
    c. Presented by the Rev. Dan Greatorex, Feb. 24, 1866.
    d. Deposited, June 7, 1866.

Genus STURNIA.

146. *Sturnia malabarica* (Gm.).　Malabar Mynah.
Hab. Hindostan.

    a–c. Purchased, May 30, 1862.

Genus STURNOPASTOR.

147. *Sturnopastor contra* (Linn.).　Pied Mynah.
Hab. India.

    a. Purchased, June 10, 1863.

Genus GRACUPICA.

148. *Gracupica nigricollis,* Paykull.　Black-necked Grac
Hab. China.

    a, b. Purchased, May 22, 1866.

Genus GRACULA.

149. *Gracula intermedia,* Hay.　Indian Mynah.
Hab. Hindostan.

    a. Presented by M. J. Harpley, Esq., October 29, 1859.
    b. Purchased, June 28, 1862.
    c. Purchased, May 8, 1865.
    d. Purchased, June 14, 1866.

150. *Gracula religiosa*, Linn.   Small Hill-Mynah.
Hab. South India.

> *a.* Presented by George Graham, Esq., Nov. 12, 1866.

Genus PTILONORHYNCHUS.

151. *Ptilonorhynchus holosericeus*, Kuhl.   Bower-bird.
Hab. New South Wales.

> *a.* Male.   Presented by Thomas Walker, Esq., F.Z.S., Dec. 24,
> 1857.
> *b.* Purchased, Jan. 4, 1865.
> *c.* Purchased, Jan. 12, 1865.
> *d.* Male ; *e.* Female.   Presented by the Acclimatization So-
> ciety of Sydney, April 19, 1866.
> *f.* Purchased, May 9, 1866.

Family PARADISEIDÆ.

Genus PARADISEA.

152. *Paradisea papuana*, Shaw.   Lesser Bird of Paradise.
Hab. New Guinea.

> *a.* Male.   Purchased for the Society at Singapore, and brought
> home by A. R. Wallace, Esq., F.Z.S., April 1, 1862.

Family CORVIDÆ.

Genus BARITA.

153. *Barita destructor*, Temm.   Long-billed Butcher-bird.
Hab. New Holland.

> *a.* Presented by Charles Clifton, Esq., F.Z.S., May 4, 1863.
> *b.* Deposited, July 21, 1866.

Genus GYMNORHINA.

154. *Gymnorhina leuconota*, Gould.   White-backed Piping
Crow.
Hab. South Australia.

> *a, b.* Presented by Dr. Mueller, C.M.Z.S., Aug. 19, 1862.
> From Melbourne.

c–l. Deposited, 1863.
m. Deposited, July 20, 1866.
n. Presented by C. C. Fuller, Esq., Dec. 21, 1866.

155. *Gymnorhina organica,* Gould.   Tasmanian Piping Crow.
Hab. Tasmania.

a, b. Presented by Dr. Mueller, C.M.Z.S., Oct. 26, 1863.

Genus STREPERA.

156. *Strepera anaphonensis* (Temm.).   Grey Crow-Shrike.
Hab. Australia.

a. Purchased, Aug. 16, 1864.
b, c. Presented by Miss Mary J. Mayo, May 5, 1865.
d. Received, April 4, 1866.

157. *Strepera arguta,* Gould.   Hill Crow-Shrike.
Hab. Tasmania.

a. Presented by Miss Mary J. Mayo, May 5, 1865.

Genus PICA.

158. *Pica caudata,* Flem.   Common Magpie.
Hab. British Islands.

a. Presented by Miss Perry, Dec. 17, 1858.
b. Presented by William Russell, Esq., F.Z.S., Sept. 15, 1862.
c. Presented by the Rev. Thomas Gregory, Dec. 24, 1866.

159. *Pica sericea,* Gould.   Chinese Magpie.
Hab. China.

a. Purchased, Jan. 5, 1866.

Genus DENDROCITTA.

160. *Dendrocitta vagabunda* (Lath.).   Wandering Tree-Pie.
Hab. India.

a. Purchased, Jan. 29, 1866.

Genus FREGILUS.

61. *Fregilus graculus*, Cuv.   Cornish Chough.
    Hab. British Islands.

    a, b. Purchased, June 9, 1857.

Genus CORCORAX.

62. *Corcorax leucoptera* (Temm.).   White-winged Chough.
    Hab. Australia.

    a. Purchased, May 9, 1866.

Genus CYANOCITTA.

63. *Cyanocitta cristata* (Linn.).   Blue Jay.
    Hab. North America.

    a. Purchased, July 14, 1855.
    b, c. Purchased, 1859.

Genus CYANOCORAX.

64. *Cyanocorax cyanopogon*, Max.   Blue-bearded Jay.
    Hab. Para.

    a, b. Purchased, March 8, 1864.

65. *Cyanocorax pileatus* (Temm.).   Pileated Jay.
    Hab. La Plata.

    a. Presented by Mrs. Laird Warren, Feb. 28, 1865.

Genus UROCISSA.

66. *Urocissa sinensis* (Bodd.).   Chinese Magpie.
    Hab. China.

    a. Purchased, Feb. 19, 1861.

67. *Urocissa magnirostris*, Blyth.   Siamese Magpie.
    Hab. Siam.

    a. Purchased, Aug. 14, 1862.

Genus GARRULUS.

168. *Garrulus glandarius* (Linn.). Jay.
Hab. British Islands.

a, b. Purchased, June 12, 1863.
c, d. Presented, Sept. 3, 1866.

Genus CORVUS.

169. *Corvus corax*, Linn. Raven.
Hab. British Islands.

a. Presented by the Rev. W. Wilmott, Jan. 15, 1862.
b, c. Presented by W. R. Taunton, Esq., March 14, 1865.
d. Presented by the Rt. Hon. John Evelyn Denison, M.P.,
   Speaker of the House of Commons, Feb. 2, 1866.

170. *Corvus carnivorus*, Bartr. American Raven.
Hab. North America.

a. Presented by Capt. David Herd, H.B.C.S., C.M.Z.S., Oct. 7,
   1862. From the Hudson's Bay Territory.

171. *Corvus australis*, Gould. Australian Crow.
Hab. South Australia.

a. Presented by Miss Mary J. Mayo, May 5, 1865.

172. *Corvus cornix*, Linn. Hooded Crow.
Hab. British Islands.

a, b. Purchased, Feb. 11, 1862.
c. Presented by Henry Thurnall, Esq., July 25, 1866.

173. *Corvus scapulatus*, Daud. White-necked Crow.
Hab. Africa.

a. Purchased, Sept. 9, 1865.
b–d. Purchased, Aug. 30, 1866.

174. *Corvus collaris*, Strickl. White-naped Jackdaw.
Hab. Macedonia.

a. Presented by William Russell, Esq., F.Z.S., Sept. 15, 1862.

Genus Nucifraga.

175. *Nucifraga caryocatactes*, Briss. Nutcracker.
Hab. Europe.

a. Purchased, Oct. 11, 1864.
b. Received in exchange, Sept. 27, 1865.

Genus Ptilostomus.

176. *Ptilostomus senegalensis* (Briss.). Piapec.
Hab. West Africa.

a. Purchased, Aug. 19, 1865.

Family COTINGIDÆ.

Genus Chasmorhynchus.

177. *Chasmorhynchus nudicollis* (Vieill.). Naked-throated Bell-bird.
Hab. Brazil.

a. Male. Purchased, May 22, 1866.
b. Male. Purchased, Sept. 6, 1866.

Genus Rupicola.

178. *Rupicola crocea*, Vieill. Demerarau Cock-of-the-Rock.
Hab. Demerara.

a, b. Presented by J. Lucie Smith, Esq., July 4, 1866.

Order PICARIÆ.

Family PODARGIDÆ.

Genus Podargus.

179. *Podargus cuvieri*, Vig. et Horsf. Cuvier's Podargus.
Hab. Tasmania.

a. Purchased, Jan. 14, 1862.

    b. Presented by the Acclimatization Society of Victoria. Jan. 4, 1865.

    c. Presented by Dr. Mueller, C.M.Z.S., March 10, 1866.

Family MOMOTIDÆ.

Genus MOMOTUS.

180. *Momotus subrufescens,* Sclater.   Carthagenian Motmot. *Hab.* Carthagena.

    a. Purchased, July 17, 1860.

Family ALCEDINIDÆ.

Genus ALCEDO.

181. *Alcedo ispida,* Linn.   Kingfisher. *Hab.* British Islands.

    a. Presented by P. Symonds, Esq., June 11, 1863.
    b. Presented by J. C. Cuming, Esq., May 22, 1861.
    c. Presented by A. Yates, Esq., July 12, 1864.
    d. Purchased, June 24, 1865.
    e. Purchased, July 16, 1865.
    f. Presented by William Thompson, Esq., Sept. 20, 1865.
    g, h. Presented by the Rev. J. Climenson, June 6, 1866.
    i, j. Presented by Mrs. Falkner, June 7, 1866.

Genus DACELO.

182. *Dacelo gigantea* (Lath.).   Laughing Kingfisher. *Hab.* Australia.

    a–c. Purchased, June 4, 1856.
    d. Presented by Dr. Mueller of Melbourne, C.M.Z.S., May 11. 1860.
    e. Presented by Capt. Watson, July 4, 1861.
    f–i. Presented by L. Mackinnon, Esq., April 19, 1864.
    j, k. Presented by F. W. Draper, Esq., April 7, 1865.
    l. Presented by Miss Wildash, June 23, 1865.
    m. Deposited, July 20, 1866.

Family BUCEROTIDÆ.

Genus BUCEROS.

183. *Buceros bicornis*, Linn.　Concave-casqued Hornbill.
Hab. India.

    a. Male.　Presented by the Babu Rajendra Mullick, C.M.Z.S., July 25, 1864.
    b. Female.　Presented by William Dunn, Esq., C.M.Z.S., July 25, 1864.
    c. Female.　Deposited, July 23,1865, and afterwards purchased.

184. *Buceros rhinoceros*, Linn.　Rhinoceros Hornbill.
Hab. Malay Peninsula.

    a. Presented by the Babu Rajendra Mullick, C.M.Z.S., July 25, 1864.

185. *Buceros albirostris*, Shaw.　White-billed Hornbill.
Hab. India.

    a. Presented by the Babu Rajendra Mullick, C.M.Z.S., March 31, 1863.

186. *Buceros convexus*, Temm.　Malayan Pied Hornbill.
Hab. Malacca.

    a. Purchased, Jan. 1, 1866.

Genus TOCCUS.

187. *Toccus erythrorhynchus*, Temm.　Red-billed Hornbill.
Hab. Gambia.

    a. Purchased, May 19, 1865.

Genus BUCORVUS.

188. *Bucorvus abyssinicus* (Gm.).　Ground-Hornbill.
Hab. West Africa.

    a. Female.　Purchased, Sept. 9, 1865.
    b, c. Females.　Purchased, Aug. 30, 1866.

Family UPUPIDÆ.

Genus UPUPA.

189. *Upupa epops*, Linn.   Hoopoe.
Hab. Europe.

  a. Presented by F. Campbell, Esq., July 2, 1865.
  b, c. Purchased, Aug. 13, 1866.

Family CUCULIDÆ.

Genus CUCULUS.

190. *Cuculus canorus*, Linn.   Common Cuckoo.
Hab. British Islands.

  a. Presented by R. F. Lascelles, Esq., Jun., Sept. 11, 1863.
  b. Presented by J. Currie, Esq., Nov. 15, 1863.
  c. Presented by W. Barrington d'Almeida, Esq., Sept. 19, 1865.
  d. Presented by E. C. Hampton, Esq., Oct. 22, 1866.

Genus GUIRA.

191. *Guira piririgua* (Vieill.).   Guira Cuckoo.
Hab. Para.

  a–d. Purchased, March 8, 1864.

Genus EUDYNAMYS.

192. *Eudynamys orientalis* (Linn.).   Black Cuckoo.
Hab. India.

  a. Male ; *b*. Female.   Presented by the Babu Rajendra Mullick, C.M.Z.S., July 25, 1864.

Family MUSOPHAGIDÆ.

Genus MUSOPHAGA.

193. *Musophaga violacea*, Isert.   Violaceous Plantain-cutter.
Hab. West Africa.

  a. Deposited, April 24. 1863.

b. Purchased, Dec. 13, 1866.
c. Young. Purchased, Dec. 13, 1866.

Genus SCHIZORHIS.

194. *Schizorhis africana* (Lath.). Variegated Touracou.
Hab. West Africa.

 a. Purchased, Dec. 3, 1863.
 b. Purchased, March 16, 1866.

Genus CORYTHAIX.

195. *Corythaix buffonii* (Vieill.). Buffon's Touracou.
Hab. West Africa.

 a. Presented by Russell M. Gordon, Esq., F.Z.S., Oct. 10, 1862.

196. *Corythaix persa* (Linn.). Senegal Touracou.
Hab. West Africa.

 a. Presented by Admiral Sir H. Keppel, K.C.B., Aug. 14, 1861.
 b. Presented by Lieut. A. H. Webb, R.N., July 28, 1858.

197. *Corythaix macrorhyncha*, Fraser. Great-billed Touracou.
Hab. West Africa.

 a. Purchased, Dec. 28, 1865.
 b. Purchased, Dec. 13, 1866.

Family RAMPHASTIDÆ.

Genus RAMPHASTOS.

198. *Ramphastos toco*, Gm. Toco Toucan.
Hab. Guiana.

 a. Presented by F. Anderson, Esq., Nov. 18, 1863.

199. *Ramphastos ariel*, Vig. Ariel Toucan.
Hab. Brazil.

 a. Purchased, July 4, 1859.

200. *Ramphastos carinatus,* Swains.   Sulphur-breasted Toucan.
Hab. Mexico.

  a. Purchased, March 22, 1860.
  b. Purchased, Oct. 15, 1866.

Family BUCCONIDÆ.

Genus MEGALÆMA.

201. *Megalæma asiatica* (Lath.).   Blue-cheeked Barbet.
Hab. India.

  a. Purchased, Sept. 4, 1866.

Family PICIDÆ.

Genus PICUS.

202. *Picus major,* Linn.   Greater Spotted Woodpecker.
Hab. British Islands.

  a. Purchased, June 12, 1863.
  b, c. Presented by S. C. Hincks, Esq., June 23, 1865.
  d. Presented by H. S. H. Prince Edward of Saxe-Weimar, Aug. 3, 1865.

Genus COLAPTES.

203. *Colaptes auratus* (Linn.).   Golden-winged Woodpecker.
Hab. North America.

  a. Purchased, March 3, 1864.
  b. Purchased, Aug. 3, 1866.

Order PREHENSORES.

Family PSITTACIDÆ.

Subfamily PSITTACINÆ.

a. Americanæ.

Genus ARA.

204. *Ara maracana* (Vieill.).   Illiger's Maccaw.
Hab. Brazil.

a. Purchased, June 7, 1861.

205. *Ara glauca*, Vieill.   Glaucous Maccaw.
Hab. Brazil.

a. Purchased, June 1860.

206. *Ara ararauna* (Linn.).   Blue-and-Yellow Maccaw.
Hab. South America.

a. Purchased, Aug. 24, 1859.
b. Presented by R. W. Keate, Esq., F.Z.S., Aug. 9, 1862.
c. Presented by Mrs. Heathcote, July 14, 1862.
d. Deposited, Sept. 4, 1866.

207. *Ara macao* (Linn.).   Red-and-Blue Maccaw.
Hab. Central America.

a. Deposited, Nov. 7, 1859.
b. Deposited, July 12, 1866.

208. *Ara chloroptera*, Gray.   Red-and-Yellow Maccaw.
Hab. South America.

a. Deposited, May 24, 1861.
b, c. Presented by the Prince de Joinville, Oct. 13, 1863.
d. Presented by J. Aird, Esq., Jan. 14, 1865.
e. Presented by Charles Butler, Esq., Jan. 18, 1865.
f. Deposited, Jan. 18, 1865.

209. *Ara militaris* (Linn.).   Lesser Military Maccaw.
Hab. South America.

a. Purchased, Jan. 24, 1864.
b. Purchased, July 5, 1866.

210. *Ara ambigua* (Bechst.). Larger Military Maccaw.
   Hab. Mexico.

   a. Purchased, Aug. 23, 1866.

Genus CONURUS.

211. *Conurus carolinensis* (Linn.). Carolina Conure.
   Hab. North America.

   a. Purchased, Aug. 23, 1860.

212. *Conurus holochlorus,* Sclater. Mexican Conure.
   Hab. Mexico.

   a, b. Purchased, Nov. 11, 1862.

213. *Conurus erythrogenys* (Less.). Red-masked Conure.
   Hab. Ecuador.

   a. Purchased, 1854.

214. *Conurus solstitialis* (Linn.). Yellow Conure.
   Hab. Brazil.

   a. Purchased, April 25, 1862.

215. *Conurus hæmorrhous,* Spix. Blue-crowned Conure.
   Hab. Brazil.

   a. Purchased, April 12, 1864.

216. *Conurus aureus* (Gm.). Golden-crowned Conure.
   Hab. Brazil.

   a, b. Presented by Miss Langford, Dec. 22, 1862.
   c, d. Presented by Carl A. Schroder, Esq., July 29, 1865.
      From Para.
   e, f. Presented by Miss Vawser, Nov. 9, 1865.

217. *Conurus xantholæmus,* Sclater. St. Thomas's Conure.
   Hab. St. Thomas, West Indies.

   a, b. Purchased, Sept. 12, 1865.

218. *Conurus chrysogenys*, Mass. & Souancé. Yellow-cheeked Conure.
   Hab. Venezuela.

   a. Purchased, 1853.

219. *Conurus æruginosus* (Linn.). Brown-throated Conure.
   Hab. South America.

   a, b. Presented by Mrs. C. Vinall, Sept. 21, 1866.

220. *Conurus cactorum* (Max.). Cactus-Conure.
   Hab. Bahia.

   a. Purchased, Aug. 29, 1862.

221. *Conurus monachus* (Bodd.). Grey-breasted Conure.
   Hab. Monte Video

   a. Presented by Mrs. Malcolm, Aug. 5, 1859.

222. *Conurus smaragdinus* (Gm.). Chilian Conure.
   Hab. Chili.

   a. Presented by T. O. Simpson, Esq., R.N., Aug. 10, 1866.

223. *Conurus tiriacula* (Bodd.). All-green Conure.
   Hab. South America.

   a, b. Presented by Lady Gilbert, June 17, 1862.

224. *Conurus virescens* (Gm.). Yellow-winged Conure.
   Hab. Brazil.

   a. Purchased, Oct. 20, 1862.

Genus BROTOGERYS.

225. *Brotogerys pyrrhopterus* (Lath.). Orange-winged Parrakeet.
   Hab. Western Ecuador.

   a, b. Purchased, Aug. 29, 1862.

226. *Brotogerys tui* (Gm.). Golden-headed Parrakeet.
   Hab. South America.

   a. Purchased, Oct. 20, 1862.

Genus PSITTACULA.

227. *Psittacula passerina* (Linn.).　Passerine Parrakeet.
Hab. Tropical America.

a. Presented by — Davis, Esq., May 24, 1858.

Genus CAÏCA.

228. *Caïca melanocephala* (Linn.).　Black-headed Parrot.
Hab. Demerara.

a. Purchased, 1855.
b, c. Purchased, Aug. 14, 1866.

Genus PIONUS.

229. *Pionus maximiliani* (Kuhl).　Maximilian's Parrot.
Hab. Brazil.

a. Purchased, July 31, 1862.

230. *Pionus senilis* (Spix).　White-headed Parrot.
Hab. Mexico.

a. Purchased, Oct. 6, 1862.

Genus DEROPTYUS.

231. *Deroptyus accipitrinus* (Linn.).　Hawk-headed Parro
Hab. Brazil.

a. Presented by G. Dennis, Esq., July 19, 1856.

Genus CHRYSOTIS.

232. *Chrysotis levaillantii*, Gray.　Levaillant's Amazon.
Hab. Mexico.

a. Purchased, Aug. 3, 1859.

233. *Chrysotis auripalliata* (Less.).　Golden-naped Amazo
Hab. Central America.

a. Purchased, 1844.
b. Presented by Lady Ousley, March 24, 1866.

234. *Chrysotis sallæi,* Sclater.   Sallé's Amazon.
Hab. St. Domingo.

    a. Presented by William Russell, Esq., F.Z.S., Sept. 15, 1862.

235. *Chrysotis farinosa* (Bodd.).   Mealy Amazon.
Hab. South America.

    a. Deposited, March 18, 1863.
    b. Presented by J. Aveling, Esq., Aug. 8, 1863.

236. *Chrysotis dufresniana* (Kuhl).   Dufresne's Amazon.
Hab. Brazil.

    a. Purchased, Jan. 10, 1863.

237. *Chrysotis viridigenalis,* Cass.   Green-cheeked Amazon.
Hab. Mexico.

    a. Purchased, Jan. 10, 1863.

238. *Chrysotis augustus* (Vig.).   Imperial Parrot.
Hab. Dominica.

    a. Presented by P. N. Bernard, Esq., May 12, 1865.

239. *Chrysotis amazonica* (Gm.).   Blue-fronted Amazon.
Hab. Pernambuco.

    a. Deposited, May 13, 1864.
    b. Presented by Mrs. Sotheby, July 22, 1865.

240. *Chrysotis festiva* (Linn.).   Festive Amazon.
Hab. Para.

    a. Purchased, Nov. 16, 1865.

b. Africanæ.

Genus PSITTACUS.

241. *Psittacus erithacus,* Linn.   Grey Parrot.
Hab. West Africa.

    a. Purchased, 1854.
    b. Deposited, Sept. 15, 1862.
    c. Received, March 4, 1865.
    d, e. Deposited, Aug. 16, 1865.

f. Deposited, Oct. 31, 1865.
g. Deposited, May 10, 1866.

242. *Psittacus timneh*, Fraser.   Timneh Parrot.
Hab. Sierra Leone.

a. Purchased, Feb. 19, 1861.

Genus CORACOPSIS.

243. *Coracopsis vasa* (Linn.).   Greater Vasa Parrakeet.
Hab. Madagascar.

a. Presented by the late Charles Telfair, Esq., C.M.Z.S
July 25, 1830.
b. Presented by Mrs. Moon, May 11, 1866.

244. *Coracopsis nigra* (Linn.).   Lesser Vasa Parrakeet.
Hab. Madagascar.

a. Purchased, Jan. 17, 1857.

Genus AGAPORNIS.

245. *Agapornis cana* (Gm.).   Grey-headed Parrakeet.
Hab. Madagascar.

a. Male; *b.* Female.   Purchased, March 17, 1860.

246. *Agapornis pullaria* (Linn.).   Love-bird Parrakeet.
Hab. West Africa.

a. Male; *b.* Female.   Purchased, April 17, 1863.

247. *Agapornis roseicollis* (Vieill.).   Rosy-faced Parrakeet.
Hab. South Africa.

a. Purchased, July 10, 1862.
b. c. Purchased, May 22, 1866.

Genus PŒOCEPHALUS.

248. *Pœocephalus meyerii* (Rüpp.).   Meyer's Parrot.
Hab. East Africa.

a. Purchased, Jan. 18, 1855.

249. *Pæocephalus senegalensis* (Linn.). Senegal Parrot
Hab. West Africa.

    a. Presented by Mrs. Clark, Oct. 11, 1853.
    c. Presented by Miss M. D. Du Carn, April 8, 1865.
    d. Purchased, Sept. 13, 1865.
    e, f. Purchased, Aug. 29, 1862.

250. *Pæocephalus levaillantii* (Lath.). Levaillant's Parrot.
Hab. South Africa.

    a. Presented by Mrs. Jesse, June 14, 1853.
    b. Deposited, June 26, 1866.

251. *Pæocephalus gulielmi* (Jard.). Jardine's Parrot.
Hab. West Africa.

    a. Deposited, Sept. 15, 1862.

c. Asiaticæ.

Genus PALÆORNIS.

252. *Palæornis javanica* (Osb.). Javan Parrakeet.
Hab. Java.

    a. Purchased, Oct. 20, 1859.
    b. Purchased, March 20, 1862.
    c. Deposited, Feb. 27, 1865.

253. *Palæornis luciani*, Verr. Red-cheeked Parrakeet.
Hab. East Indies.

    a. Purchased, March 11, 1857.

254. *Palæornis bengalensis* (Linn.). Blossom-headed Parrakeet.
Hab. Hindostan.

    a. Deposited, April 4, 1862.
    b. Purchased, April 15, 1862.
    c, d. Received, Jan. 14, 1865.

255. *Palæornis columboides*, Vig. Malabar Parrakeet.
Hab. South India.

    a. Purchased, June 2, 1852.

256. *Palæornis torquata* (Linn.).   Ring-necked Parrakeet.
Hab. India.

    a. Female.   Presented by Mrs. James Part, Jan. 31, 1863.
    b. Deposited, 1862.
    c. Deposited, Oct. 29, 1863.
    d. Presented by Miss Bushby, June 8, 1865.
    e. Presented by — Toplis, Esq., Aug. 30, 1865.
    f. Deposited, Jan. 29, 1866.
    g. Presented by — Verzey, Esq., Dec. 11, 1866.

257. *Palæornis alexandri* (Linn.).   Alexandrine Parrakeet.
Hab. Hindostan.

    a. Female.   Presented by Richard Tress, Esq., F.Z.S., Oct.
       22, 1855.
    b. Presented by Mrs. Roper, Nov. 3, 1862.
    c. Deposited, Aug. 18, 1863.
    d. Presented by Mrs. Synd, Aug. 4, 1864.

258. *Palæornis docilis*, Gray.   Rose-ringed Parrakeet.
Hab. West Africa.

    a. Presented by George Dann, Esq., Sept. 13, 1861.

259. *Palæornis malaccensis*, Vig.   Malaccan Parrakeet.
Hab. Malacca.

    a, b. Purchased, Feb. 29, 1864.

Genus PSITTINUS.

260. *Psittinus malaccensis* (Lath.).   Blue-rumped Parrakee .
Hab. Malacca.

    a. Female.   Purchased, Jan. 29, 1866.

Genus TANYGNATHUS.

261. *Tanygnathus macrorhynchus* (Linn.).   Great-billed
    Parrakeet.
Hab. Gilolo and Ceram.

    a. Purchased, April 20, 1856.

262. *Tanygnathus muelleri* (Temm.). Mueller's Great-billed Parrakeet.
Hab. Celebes.

 a. Presented by the Babu Rajendra Mullick, C.M.Z.S., July 14, 1857.

Genus ECLECTUS.

263. *Eclectus linnæi*, Wagl. Linnean Eclectus.
Hab. Moluccas.

 a. Presented by George Macleay, Esq., F.Z.S., Aug. 19, 1859.

264. *Eclectus grandis* (Gm.). Grand Eclectus.
Hab. Gilolo.

 a, b. Purchased, June 8, 1865.
 c. Purchased, Aug. 30, 1865.

265. *Eclectus magnus* (Scop.). Red-sided Eclectus.
Hab. Moluccas.

 a. Purchased, April 11, 1864.
 b, c. Purchased, Jan. 17, 1865.

266. *Eclectus westermanni*, Bp. Westermann's Eclectus.
Hab. Moluccas.

 a. Purchased, Nov. 23, 1865.

Genus LORICULUS.

267. *Loriculus sclateri*, Wall. Sclater's Hanging Parrakeet.
Hab. Sulla Islands.

 a. Purchased, Nov. 23, 1865.

Subfamily PLATYCERCINÆ.

Genus LATHAMUS.

268. *Lathamus discolor* (Shaw). Swift Parrakeet.
Hab. Van Diemen's Land.

 a, b. Purchased, May 17, 1863.

Genus PEZOPORUS.

269. *Pezoporus formosus* (Lath.). Green Ground-Parrakeet.
Hab. Australia.

a. Purchased, May 5, 1865.

Genus MELOPSITTACUS.

270. *Melopsittacus undulatus* (Shaw). Undulated Grass-
Parrakeet.
Hab. Australia.

a–c. Deposited, 1862.
d–f. Deposited, Nov. 14, 1863.
g–l. Bred in the Gardens, 1863.
m, n. Deposited, Sept. 16, 1863.
o–q. Bred in the Gardens, Dec. 12, 1864.
r. Purchased, Sept. 16, 1865.
s. Presented by Miss Boyle, Dec. 14, 1865.

Genus EUPHEMA.

271. *Euphema pulchella* (Shaw). Turquoisine Parrakeet.
Hab. New South Wales.

a. Male: *b.* Female. Bred in the Gardens, 1860.
c. Female. Deposited, May 19, 1863.
d–f. Bred in the Gardens, 1863.
g, h. Deposited, March 7, 1865.
i, j. Bred in the Gardens, June 2, 1865.
k, l. Bred in the Gardens, July 27, 1865.
m, n. Bred in the Gardens, Aug. 29, 1865.
o–q. Bred in the Gardens, Sept. 26, 1865.
r, s. Bred in the Gardens, Feb. 1, 1866.

272. *Euphema elegans,* Gould. Elegant Grass-Parrakeet.
Hab. South Australia.

a. Presented by William Russell, Esq., F.Z.S., Sept. 15, 1862.

Genus CALOPSITTA.

273. *Calopsitta novæ-hollandiæ* (Gm.). Crested Ground-
Parrakeet.
Hab. Australia.

a. Presented by T. W. Nunn, Esq., March 13, 1862.
b. Male: *c.* Female. Received in exchange, April 17, 1862.

d. Male; *e.* Female. Deposited, Nov. 14, 1863.
f. Female. Deposited, Oct. 28, 1863.
g. Bred in the Gardens, Aug. 12, 1863.
h. Bred in the Gardens, Dec. 12, 1864.
i. Deposited, July 7, 1865.
j. Bred in the Gardens, Feb. 1, 1866.

Genus PSEPHOTUS.

274. *Psephotus multicolor* (Brown). Many-coloured Parrakeet.
    Hab. Australia.

    a. Male; *b.* Female. Purchased, April 25, 1862.

275. *Psephotus hæmatogaster*, Gould. Blue-bonnet Parrakeet.
    Hab. Australia.

    a. Purchased, April 25, 1862.

276. *Psephotus hæmatonotus*, Gould. Blood-rumped Parrakeet.
    Hab. Australia.

    a. Male. Purchased, May 29, 1862.
    b. Female. Bred in the Gardens.

277. *Psephotus pulcherrimus*, Gould. Beautiful Parrakeet.
    Hab. Australia.

    a. Male; *b.* Female. Purchased, March 28, 1866.

Genus PLATYCERCUS.

278. *Platycercus palliceps*, Vig. Pale-headed Parrakeet.
    Hab. Moreton Bay, North-east Australia.

    a, b. Presented by Sir John Cathcart, Aug. 21, 1863.

279. *Platycercus melanurus* (Vig.). Black-tailed Parrakeet.
    Hab. Australia.

    a, b. Purchased, March 15, 1864.
    c, d. Received, Jan. 14, 1865.

280. *Platycercus pileatus,* Vig.   Pileated Parrakeet.
     Hab. Australia.

     a. Purchased, May 27, 1854.
     b, c. Purchased, July 20, 1865.

281. *Platycercus eximius* (Shaw).   Rose-Hill Parrakeet.
     Hab. New South Wales.

     a. Presented by G. H. Parkinson, Esq., Nov. 23, 1861.
     b–g. Deposited, July 20, 1866.

282. *Platycercus pennantii* (Lath.).   Pennant's Parrakeet.
     Hab. New South Wales.

     a, b. Presented by Mrs. Wheeler, Aug. 20, 1861.
     c. Received in exchange, July 29, 1865.

283. *Platycercus flaviventris* (Vig. et Horsf.).   Yellow-bellied
     Parrakeet.
     Hab. Tasmania.

     a. Purchased, May 8, 1860.

284. *Platycercus adelaidæ,* Gould.   Adelaide Parrakeet.
     Hab. South Australia.

     a. Deposited, Nov. 14, 1863.

285. *Platycercus icterotis* (Temm.).   Stanley Parrakeet.
     Hab. King George's Sound.

     a. Received in exchange, Nov. 15, 1864.
     b, c. Purchased, July 20, 1865.

286. *Platycercus barnardii* (Lath.).   Barnard's Parrakeet.
     Hab. South Australia.

     a. Purchased, April 25, 1853.
     b, c. Deposited, Nov. 14, 1863.

287. *Platycercus zonarius* (Shaw).   Bauer's Parrakeet.
     Hab. South Australia.

     a. Presented by Sir John Cathcart, Aug. 21, 1863.

288. *Platycercus semitorquatus* (Quoy et Gaim.).   Yellow-collared Parrakeet.
    Hab. West Australia.

    a. Presented by F. J. Rooper, Esq., F.Z.S., July 8, 1862.

289. *Platycercus auriceps,* Vig.   Golden-headed Parrakeet.
    Hab. New Zealand.

    a. Purchased, Nov. 24, 1865.

290. *Platycercus novæ-zealandiæ* (Gm.).   New Zealand Parrakeet.
    Hab. New Zealand.

    a. Purchased, April 7, 1864.

Genus APROSMICTUS.

291. *Aprosmictus erythropterus* (Lath.).   Red-winged Parrakeet.
    Hab. Australia.

    a. Female.   Presented by R. Marshall, Esq., F.Z.S., March 30, 1861.
    b. Male.   Purchased, June 7, 1861.

292: *Aprosmictus scapulatus* (Bechst.).   King Parrakeet.
    Hab. New South Wales.

    a. Male.   Presented by Sir John Cartheart, Aug. 21, 1863.
    b. Female.   Purchased, Dec. 9, 1859.
    c. Female.   Presented by F. Hansard Rivington, Esq., Oct. 6, 1865.

Genus PYRRHULOPSIS.

293. *Pyrrhulopsis splendens,* Cass.   Shining Parrakeet.
    Hab. Fiji Islands.

    a. Purchased, June 27, 1864.
    b. Presented by Charles Moore, Esq., April 19, 1866.

Subfamily CACATUINÆ.

Genus LICMETIS.

294. *Licmetis tenuirostris* (Wagl.). Slender-billed Cockatoo.
Hab. South Australia.

a. Received in exchange, May 17, 1847.
b. Received in exchange, April 11, 1866.

295. *Licmetis pastinator,* Gould. Western Slender-billed Cockatoo.
Hab. Western Australia.

a. Presented by Edgar Ray, Esq., Sept. 22, 1858.

Genus CACATUA.

296. *Cacatua moluccensis* (Gm.). Rose-crested Cockatoo.
Hab. Moluccas.

a. Deposited, Aug. 30, 1855.
b. Presented by — Mostyn, Esq., Oct. 23, 1862.
c. Presented by Robert Drummond, Esq., F.Z.S., March 19, 1863.
d. Deposited, July 23, 1863.
e. Presented by Mrs. Moss King, Nov. 13, 1863.
f. Received in exchange, April 9, 1863.
g. Deposited, Feb. 25, 1865.

297. *Cacatua ophthalmica,* Sclater. Blue-eyed Cockatoo.
Hab. Salomon Islands.

a. Purchased, April 25, 1862.
b, c. Purchased, May 11, 1865. Specimens described and figured P.Z.S. 1862, pl. xiv., p. 141 as *Cacatua ducorpsii,* and P.Z.S. 1864, p. 187.

298. *Cacatua cristata* (Linn.). Greater White-crested Cockatoo.
Hab. Moluccas.

a. Purchased, July 19, 1861.

299. *Cacatua galerita* (Lath.). Greater Sulphur-crested Cockatoo.
Hab. Australia.

a. Presented by Richard Tress, Esq., F.Z.S., July 8, 1860.

b. Deposited, Dec. 15, 1860.
c. Presented by — Thompson, Esq., Aug. 9, 1865.

300. *Cacatua triton,* Temm.   Triton Cockatoo.
Hab. New Guinea.

a. Purchased, 1860.

301. *Cacatua sulphurea* (Gm.).   Lesser Sulphur-crested Cockatoo.
Hab. Moluccas.

a. Presented by H. Wickens, Esq., April 15, 1850.
b. Deposited, 1860.
c. Deposited, March 17, 1865.
d. Deposited, April 13, 1866.

302. *Cacatua citrino-cristata,* Fraser.   Citron-crested Cockatoo.
Hab. Timor Laut.

a. Presented by Miss Julia Fox, Dec. 22, 1855.
b. Deposited, July 17, 1866.

303. *Cacatua leadbeateri* (Vig.).   Leadbeater's Cockatoo.
Hab. Australia.

a. Presented by Lady Eleanor Cathcart, Nov. 21, 1854.
b. Deposited, Nov. 9, 1863.
c–e. Deposited, April 8, 1863.
f, g. Presented by Dr. Mueller, C.M.Z.S., May 5, 1865.
h. Deposited, Oct. 31, 1865.
i. Presented by Mrs. James M. Napier, Nov 23, 1865.

304. *Cacatua sanguinea,* Gould.   Blood-stained Cockatoo.
Hab. North Australia.

a. Purchased, Nov. 23, 1865.

305. *Cacatua philippinarum* (Gm.).   Red-vented Cockatoo.
Hab. Philippine Islands.

a–c. Purchased, June 8, 1865.
d. Purchased, Nov. 23, 1865.

306. *Cacatua ducorpsii,* Hombr. et Jacq. Ducorps's Cockatoo.
Hab. Salomon Islands.

a. Purchased, 1858.
b. Purchased, July 31, 1862.   Specimen figured P. Z. S. 1864,
   pl. xvii.

307. *Cacatua roseicapilla,* Vieill. Roseate Cockatoo.
Hab. Australia.

a. Presented by Lady Rolle, F.Z.S., May 2, 1843.
b. Deposited, July 12, 1861.
c. Deposited, Aug. 19, 1859.
d. Deposited, June 7, 1865.
e. Deposited, March 10, 1866.

Genus CALLOCEPHALON.

308. *Callocephalon galeatum* (Lath.). Ganga Cockatoo.
Hab. New South Wales.

a. Male.   Purchased, Aug. 5, 1859.
b. Female.   Purchased, May 14, 1864.
c. Purchased, March 21, 1866.

Genus CALYPTORHYNCHUS.

309. *Calyptorhynchus banksii* (Lath.). Banksian Cockatoo.
Hab. New South Wales.

a. Purchased, April 25, 1862.

Subfamily NESTORINÆ.

Genus NESTOR.

310. *Nestor hypopolius* (Forst.). Ka-Ka Parrot.
Hab. New Zealand.

a. Purchased, April 15, 1863.

Subfamily LORIINÆ.

Genus LORIUS.

311. *Lorius garrulus* (Linn.). Ceram Lory.
Hab. Moluccas.

a. Male ; b. Female. Purchased, Jan. 1, 1864.
c. Deposited, Dec. 14, 1865.

Genus Eos.

312. *Eos reticulata* (Müll. et Schl.).   Blue-streaked Lory.
Hab. Timor Laut.

    a. Purchased, Jan. 14, 1862.

Genus TRICHOGLOSSUS.

313. *Trichoglossus hæmatodus* (Linn.).   Blue-faced Lorikeet.
Hab. Timor.

    a. Received in exchange, 1863.

Order ACCIPITRES.

Family CATHARTIDÆ.

Genus CATHARTES.

314. *Cathartes atratus* (Bartr.).   Black Vulture.
Hab. America.

    a, b. Presented by Dr. H. B. Holbeck, of Charleston, U.S.A.,
July 25, 1860.

315. *Cathartes aura* (Linn.).   Turkey Vulture.
Hab. America.

    a. Purchased, May 28, 1864.

316. *Cathartes californianus* (Shaw).   Californian Vulture.
Hab. Monterey, California.

    a. Presented by Colbert A. Canfield, M.D., C.M.Z.S., June 22,
1866.

Genus SARCORHAMPHUS.

317. *Sarcorhamphus gryphus* (Linn.).   Condor Vulture.
Hab. South America.

    a. Male.   Purchased, 1853.
    b. Female.   Purchased, April 29, 1856.

Genus GYPARCHUS.

318. *Gyparchus papa* (Linn.).   King Vulture.
    Hab. Tropical America.

    a. Male.   Presented by the late King of Portugal, F.Z.S.,
       Nov. 16, 1856.
    b. Female.   Presented by W. D. Christie, Esq., F.Z.S., May
       13, 1857.
    c, d. Presented by Josiah Booker, Nov. 29, 1865.   From
       Demerara.

Family VULTURIDÆ.

Genus VULTUR.

319. *Vultur cinereus*, Linn.   Cinereous Vulture.
    Hab. East Europe.

    a. Purchased, Nov. 8, 1851.
    b. Purchased, Aug. 27, 1862.

320. *Vultur auricularis*, Daud.   Sociable Vulture.
    Hab. Africa.

    a. Purchased, March 28, 1850.
    b. Purchased, March 25, 1865.   From South Africa.
    c. Deposited, March 7, 1866.

321. *Vultur calvus*, Scop.   Pondicherry Vulture.
    Hab. India.

    a, b. Purchased, May 20, 1865.

322. *Vultur occipitalis*, Burch.   Occipital Vulture.
    Hab. Africa.

    a. Purchased, Aug. 19, 1865.

Genus GYPS.

323. *Gyps fulvus* (Gm.).   Griffon Vulture.
    Hab. Europe.

    a. Presented by Cuthbert Wigham, Esq., July 26, 1856.
    b. Presented by Col. Harding, Sept. 12, 1855.   From the
       Crimea.

    c. Presented by Rees Williams, Esq., June 10, 1862.

    d–g. Presented by Lord Londesborough, F.Z.S., April 1, 1863.

    h. Presented by R. H. Holdsworth, Esq., F.Z.S., Jan. 20, 1866.
        From Portugal.

324. *Gyps rueppllii,* Bp.   Rüppell's Vulture.
    Hab. Eastern Africa.

    a. Purchased, March 16, 1865.

Genus NEOPHRON.

325. *Neophron percnopterus* (Linn.).   Egyptian Vulture.
    Hab. Africa.

    a. Presented by Edmund R. Wodehouse, Esq., April 30, 1863.
        From the Cape Colony.   Specimen noticed P. Z. S. 1865,
        p. 675.

    b. Var. *orientalis.*   Purchased, May 20, 1865.   From India.
        Specimen noticed P. Z. S. 1865, p. 675.

326. *Neophron pileatus* (Burch.).   Pileated Vulture.
    Hab. West Africa.

    a. Male; *b.* Female.   Purchased, March 16, 1865.

Genus GYPOHIERAX.

327. *Gypohierax angolensis* (Gm.).   Angola Vulture.
    Hab. West Africa.

    a. Purchased, Aug. 18, 1864.

    b. Presented by John Montague Grant, Esq., Aug. 30, 1866.

Genus GYPAËTUS.

328. *Gypaëtus meridionalis,* Bonap.   Southern Bearded Vul-
    ture.
    Hab. North-eastern Africa.

    a. Received in exchange, Oct. 4, 1851.

Genus SERPENTARIUS.

329. *Serpentarius reptilivorus,* Daud.   Secretary Vulture.
    Hab. South Africa.

    a. Received, Aug. 21, 1866.

I

Family FALCONIDÆ.

Subfamily POLYBORINÆ.

Genus MILVAGO.

330. *Milvago australis* (Gm.).   Forster's Milvago.
Hab. Falkland Islands.

   a. Purchased, Jan. 6, 1859.
   b, c. Deposited, Sept. 2, 1862.

331. *Milvago chimango* (Vieill.).   Chimango.
Hab. South America.

   a. Purchased, Nov. 8, 1851.

Genus POLYBORUS.

332. *Polyborus brasiliensis* (Gm.).   Brazilian Caracara.
Hab. South America.

   a. Presented by the late Hugh Cuming, Esq., C.M.Z.S., Oct.
     10, 1831.
   b. Purchased. Oct. 29, 1863.

Subfamily MILVINÆ.

Genus MILVUS.

333. *Milvus regalis*, Briss.   Common Kite.
Hab. Europe.

   a. Purchased, Jan. 13, 1860.
   b. Presented by the late King of Portugal, F.Z.S., Aug. 3, 1860.
   c. Presented by Henry Oakley, Esq., R.N., July 5, 1863.
   d. Presented by Howard Saunders, Esq., F.Z.S., Nov. 12,
     1866.

334. *Milvus niger*, Briss.   Black Kite.
Hab. Europe, Africa, and Asia.

   a. Purchased, Sept. 7, 1860.   From Germany.
   b. Purchased, July 12, 1861.   From Germany.
   c. Presented by G. B. Bird, Esq., May 26, 1861.   From North
     China.

d. Bred in the Gardens, June 27, 1864.
e. Presented by R. M. Young, Esq., May 17, 1865.
f. Bred in the Gardens, June 20, 1866.

335. *Milvus govinda*, Sykes.   Indian Kite.
Hab. Eastern Asia.

a. Purchased.

336. *Milvus parasiticus* (Daud.).   Parasitic Kite.
Hab. Africa.

a. Presented by the Hon. Mrs. Stuart, Dec. 16, 1861.   From
   Lake Menzaleh, Egypt.
b. Purchased, Oct. 30, 1863.   From West Africa.
c–e. Purchased, June 16, 1864.   From Egypt.

Genus ELANUS.

337. *Elanus scriptus*, Gould.   Letter-winged Kite.
Hab. King George's Sound, West Australia.

a. Presented by E. St. Jean, Esq., May 29, 1865.

Subfamily AQUILINÆ.

Genus BUTEO.

338. *Buteo vulgaris*, Bechst.   Common Buzzard.
Hab. Europe.

a. Presented by J. H. Gurney, Esq., F.Z.S., Jan. 17, 1862.
b. Presented by Professor Rolleston, F.Z.S., Aug. 11, 1864.
c. Presented by Lord Willoughby de Eresby, July 14, 1865.

339. *Buteo jacal* (Daud.).   Jackal Buzzard.
Hab. Africa.

a. Presented by H.E. Sir George Grey, K.C.B., F.Z.S., Governor
   of New Zealand, Nov. 1, 1861.   From the Cape Colony.
b. Purchased, Oct. 30, 1863.   From West Africa.
c. Presented by G. W. Baker, Esq., May 24, 1866.   From
   Natal.

340. *Buteo tachardus* (Daud.).   African Buzzard.
Hab. Africa.

a. Presented by E. L. Layard, Esq., F.Z.S., Oct. 31, 1860.
   From the Cape Colony.

341. *Buteo augur*, Rüpp.   Augur Buzzard.
Hab. West Africa.

   a, b. Purchased. April 26, 1866.

Genus ARCHIBUTEO.

342. *Archibuteo lagopus* (Gm.).   Rough-legged Buzzard.
Hab. Europe.

   a. Presented by Sir T. Fowell Buxton, Bart., F.Z.S., March 31, 1862.
   b. Purchased, June 14, 1865.

Genus PERNIS.

343. *Pernis apivorus* (Linn.).   Honey-Buzzard.
Hab. Europe.

   a. Presented by Charles Clifton, Esq., F.Z.S., Sept. 24, 1860.
   b. Presented by Percy S. Godman, Esq., C.M.Z.S., Aug. 13, 1861.   From Norway.

Genus HALIASTUR.

344. *Haliastur indus* (Bodd.).   Brahminy Kite.
Hab. South Asia.

   a. India.   Presented by the Babu Rajendra Mullick, C.M.Z.S., July 14, 1857.
   b–d. Deposited, May 20, 1865.   From India.
   e, f. Purchased, July 3, 1865.   From India.

345. *Haliastur intermedius*, Blyth.   Javan Brahminy Kite.
Hab. Java.

   a. Received in exchange, Aug. 7, 1861.

346. *Haliastur sphenurus* (Vieill.).   Whistling Kite.
Hab. Australia.

   a, b. Received, April 24, 1863.   From Queensland.

Genus HELOTARSUS.

347. *Helotarsus ecaudatus* (Daud.).   Bateleur Eagle.
Hab. Africa.

a. Presented by J. J. Monteiro, Esq., C.M.Z.S., Nov. 12, 1862.
   From Angola.
b, c. Deposited, Dec. 12, 1864.

Genus HALIAËTUS.

348. *Haliaëtus albicilla* (Linn.).   Cinereous Sea-Eagle.
Hab. Europe.

a. Presented by the Viscount Powerscourt, F.Z.S., Feb. 13, 1862.
b. Presented by — Xenos, Esq., Sept. 11, 1862.
c. Presented by J. Slowman, Esq., Nov. 6, 1862.
d. Presented by the Marquis of Salisbury, K.G., F.Z.S., Jan. 17, 1863.
e, f. Presented by H. Stafford O'Brien, Esq., Oct. 24, 1864.
g. Deposited, May 3, 1865.   From Hornet Bay, Mantchuria.
h, i. Presented by H. Buckley, Esq., Aug. 7, 1865.

349. *Haliaëtus leucocephalus* (Linn.).   White-headed Sea-Eagle.
Hab. North America.

a. Presented by Dr. E. J. Longton, Feb. 14, 1862.   Taken in the Atlantic.
b. Presented by J. Rendall, Esq., May 25, 1863.   Taken from the nest in Placentia Bay, Newfoundland.   Specimen noticed P. Z. S. 1863, p. 251, et 1865, p. 731.

350. *Haliaëtus macii* (Temm.).   Mace's Sea-Eagle.
Hab. India.

a. Presented by A. Grote, Esq., C.M.Z.S., April 20, 1863.

351. *Haliaëtus leucogaster* (Gm.).   White-bellied Sea-Eagle.
Hab. Australia.

a. Purchased, April 6, 1864.

352. *Haliaëtus vocifer* (Daud.).   Vociferous Sea-Eagle.
Hab. Africa.

a, b. Purchased, Aug. 19, 1866.
c. Presented by Edward Hooper, Esq., Jan. 18, 1866.   From Natal.

Genus GERANOAËTUS.

353. *Geranoaëtus aguia* (Temm.).   Chilian Sea-Eagle.
Hab. Chili.

a. Presented by Admiral Seymour, July 12, 1848.
b. Purchased, Aug. 18, 1864.
c. Purchased, Sept. 13, 1865.

Genus AQUILA.

354. *Aquila chrysaëtos* (Linn.).   Golden Eagle.
Hab. Europe and North America.

a. Presented by J. H. Gurney, Esq., F.Z.S., 1857.
b. Presented by Capt. David Herd, H.B.C.S., C.M.Z.S., Aug. 13,
     1858.   From the Hudson's Bay Territory.
c. Presented by Capt. David Herd, H.B.C.S., C.M.Z.S., Oct. 12,
     1863.   From the Hudson's Bay Territory.
d, e. Presented by Capt. David Herd, H.B.C.S., C.M.Z.S.,
     Oct. 27, 1864.   From the Hudson's Bay Territory.
f. On approval, May 3, 1865.
g. Deposited, Aug. 1, 1865.
h–j. Presented by Capt David Herd, H.B.C.S., C.M.Z.S.,
     Aug. 17, 1866.

355. *Aquila barthelemyi*, Jaub.   Barthelemy's Eagle.
Hab. Southern France.

a. Purchased, April 30, 1866.

356. *Aquila heliaca*, Savig.   Imperial Eagle.
Hab. Europe.

a. Presented by the late King of Portugal, F.Z.S., Aug. 30,
     1860.
b. Presented by Lord Londesborough, F.Z.S., April 1, 1862.
c, d. Purchased, Sept. 19, 1864.   From the Lower Danube.

357. *Aquila nævioïdes*, Cuv.   Tawny Eagle.
Hab. Africa.

a. Presented by E. L. Layard, Esq., F.Z.S., Oct. 31, 1860.
     From the Cape Colony.
b. Presented by W. H. Simpson, Esq., F.Z.S., 1857.   From
     Algeria.

358. *Aquila nævia* (Gm.).   Spotted Eagle.
Hab. Europe.

   a. Purchased, Oct. 17, 1863.

359. *Aquila audax* (Lath.).   Wedge-tailed Eagle.
Hab. Australia.

   a. Presented by Dr. Mueller, of Melbourne, C.M.Z.S., Jan. 8, 1861.
   b, c. Presented by Dr. Mueller, of Melbourne, C.M.Z.S., Dec. 13, 1859.
   d. Presented by Samuel Magnus, Esq., May 15, 1858.
   e. Presented by — Russell, Esq., March 26, 1863.
   f. Presented by Dr. Mueller, C.M.Z.S., May 5, 1865.

360. *Aquila pennata* (Gmel.).   Booted Eagle.
Hab. Europe.

   a. Presented by R. S. A. South, Esq., Dec. 5, 1865.

Genus SPIZAËTUS.

361. *Spizaëtus occipitalis* (Daud.).   Black Crested Eagle.
Hab. Africa.

   a. Presented by the late Edmund Gabriel, Esq., H.B.M.'s Commissioner at Loando, Angola, 4, 1860.   From Angola.
   b. Deposited, July 25, 1864.   From South Africa.
   c. Purchased, Jan. 14, 1865.
   d, e. Deposited, Nov. 22, 1865.

362. *Spizaëtus coronatus* (Linn.).   Crowned Eagle.
Hab. Senegal.

   a. Male.   Purchased, April 30, 1866.

Genus THRASAËTUS.

363. *Thrasaëtus harpyia* (Linn.).   Harpy Eagle.
Hab. South America.

   a. Purchased, April 5, 1854.
   b. Presented by the late King of Portugal, F.Z.S., Oct. 15, 1858.

120   FALCONIDÆ.

Genus HARPYHALIAËTUS.

364. *Harpyhaliaëtus coronatus* (Vieill.).   Crowned Harpy.
Hab. La Plata.

 a. Presented by Edward Wallace Goodlake, Esq., C.M.Z.S.,
    Feb. 1, 1863.

Genus HÆMATORNIS.

365. *Hæmatornis elgini*, Tytler.   Lord Elgin's Hawk.
Hab. Andaman Islands.

 a, b. Presented by A. Grote, Esq., F.Z.S., May 20, 1865.

Subfamily FALCONINÆ.

Genus FALCO.

366. *Falco grœnlandicus*, Hancock.   Greenland Falcon.
Hab. Greenland.

 a. Purchased, 1859.

367. *Falco islandicus*, Brünn.   Iceland Falcon.
Hab. Iceland.

 a. Presented by Major Delmé Radcliffe, May 4, 1862.
 b, c. Deposited, June 11, 1866.

368. *Falco peregrinus*, Linn.   Peregrine Falcon.
Hab. Europe.

 a. Presented by Thomas Fraser, Esq., Oct 22, 1862.
 b. Purchased, Nov. 8, 1862.
 c. Purchased, Dec. 3, 1863.
 d. Presented by J. M. Baragett, Esq., March 15, 1865.
 e. Presented by John Wingfield Larkin, Esq., Nov. 21, 1865.
 f. Presented by Capt. C. Griffith, Beng. Army, May 28, 1866.
 g, h. Presented by Lord F. Conyngham, July 16, 1866.   From
    Ireland.
 i-p. Deposited, July 23, 1866.
 q. Presented by Mrs. Charles Cox, Dec. 6, 1866.

369. *Falco anatum,* Bonap.   Duck Falcon.
   Hab. North America.

   a. Presented by Capt. David Herd, H.B.C.S., C.M.Z.S., Oct. 12, .
      1863.   From the Hudson's Bay Territory.
   b. Purchased, Dec. 2, 1863.

370. *Falco barbarus,* Linn.   Barbary Falcon.
   Hab. North Africa.

   a. Purchased, July 27, 1864.

371. *Falco lanarius,* Schleg.   Lanner Falcon.
   Hab. Egypt.

   a. Presented by J. H. Cochrane, Esq., F.Z.S., April 27, 1863.
   b. Deposited, July 30, 1866.
   c. Received, Sept. 10, 1866.   From Asia.

372. *Falco sacer,* Schleg.   Saker Falcon.
   Hab. Asia.

   a. Received in exchange, March 6, 1863.

373. *Falco jugger,* Gray.   Jugger Falcon.
   Hab. India.

   a. Presented by His Highness the Prince Duleep Singh, K.S.I.,
      F.Z.S., May 17, 1865.

Genus HYPOTRIORCHIS.

374. *Hypotriorchis rufigularis* (Daud.).   Rufous-throated
      Falcon.
   Hab. South America.

   a. Purchased, March 8, 1864.

375. *Hypotriorchis subbuteo* (Linn.).   Hobby.
   Hab. British Islands.

   a. Purchased, Jan. 15, 1866.
   b. Presented by J. E. Harting, Esq., F.Z.S., July 26, 1866.
   c, d. Deposited, July 30, 1866.

376. *Hypotriorchis columbarius* (Linn.).   Pigeon-Hawk.
   Hab. North America.

   a. Purchased, Sept. 18, 1866.

Genus HIERACIDEA.

377. *Hieracidea berigora* (Vig. et Horsf.).   Berigora Hawk.
Hab. Australia.

   a. Purchased, Feb. 14, 1864.
   b. Purchased, April 10, 1865.

378. *Hieracidea novæ-zealandiæ* (Gm.).   New Zealand Hawk.
Hab. New Zealand.

   a. Presented by the Acclimatization Society of Canterbury,
   New Zealand, May 9, 1866.

Genus ERYTHROPUS.

379. *Erythropus vespertinus* (Linn.).   Red-footed Falcon.
Hab. Europe.

   a. Purchased, Sept. 19, 1864.   From the Lower Danube.
   b. Male ; *c.* Female.   Received, Sept. 10, 1866.

Genus TINNUNCULUS.

380. *Tinnunculus alaudarius*, Briss.   Common Kestrel.
Hab. British Islands.

   a–c. Deposited, Oct. 29, 1862.   From Algeria.
   d, e. Presented by J. E. Harting, Esq., F.Z.S., July 27, 1866.
   f. Presented by A. Burman, Esq., Nov. 27, 1866.

Subfamily ACCIPITRINÆ.

Genus MELIERAX.

381. *Melierax polyzonus*, Rüpp.   Many-zoned Hawk.
Hab. East Africa.

   a. Purchased, Oct. 10, 1863.

Genus ASTUR.

382. *Astur palumbarius* (Linn.).   Goshawk.
Hab. Europe.

   a, b. Deposited, Oct. 11, 1861.

c, d. Deposited, July 4, 1862.
e, f. Deposited, July 2, 1866.

383. *Astur novæ-hollandiæ* (Gm.). White Goshawk.
Hab. Australia.

a. Purchased, July 12, 1859.
b. Presented by Dr. Mueller, of Melbourne, C.M.Z.S., Sept. 10, 1864.

Genus ACCIPITER.

384. *Accipiter fringillarius*, Ray. Sparrow-Hawk.
Hab. British Islands.

a. Presented by W. H. Allies, Esq., March 3, 1866.

Subfamily CIRCINÆ.

Genus CIRCUS.

385. *Circus cyaneus* (Linn.). Hen Harrier.
Hab. Europe.

a. Presented by H. P. Hensman, Esq., Dec. 21, 1863.

386. *Circus æruginosus* (Linn.). Marsh Harrier.
Hab. Europe.

a. Presented by A. S. Yates, Esq., Aug. 15, 1863.
b. Presented by J. H. Gurney, Esq., F.Z.S., Feb. 29, 1864.
c. Presented by C. A. Goring, Esq., Sept. 8, 1865.

387. *Circus cinerascens* (Mont.). Ash-coloured Harrier.
Hab. British Islands.

a. Purchased, Oct. 8, 1862.

Family STRIGIDÆ.

Genus BUBO.

388. *Bubo maximus* (Aldrov.). Great Eagle Owl.
Hab. Europe.

a. Presented by Lieut.-Gen. C. R. Fox, F.Z.S., June 22, 1858.

b. Presented by Sir John Cathcart, Bart., F.Z.S., April 28, 1859.
c. Presented by the Hon. Mrs. Stuart, Oct. 15, 1862.
d, e. Presented by Capt. Stewart, Oct. 7, 1863.

389. *Bubo virginianus* (Gm.). Virginian Eagle Owl.
Hab. North America.

a. Presented by Capt. Wishart, H.B.C.S., Oct. 12, 1858. From the Hudson's Bay Territory.
b. Presented by Capt. David Herd, H.B.C.S., C.M.Z.S., Oct. 13, 1858. From the Hudson's Bay Territory.
c. Presented by Jos. Radford, Esq., Aug. 17, 1861.
d, e. Presented by Capt. David Herd, H.B.C.S., C.M.Z.S., Oct. 12, 1863. From the Hudson's Bay Territory.
f. Presented by E. Dumaresq, Esq., Dec. 13, 1866.

390. *Bubo ascalaphus* (Sav.). Savigny's Owl.
Hab. Mogador.

a. Deposited, May 3, 1865.

391. *Bubo lacteus* (Temm.). Milky Owl.
Hab. West Africa.

a. Received in exchange, Dec. 13, 1861.
b. Purchased, Sept. 16, 1865.

392. *Bubo cinerascens,* Guérin. Grey Owl.
Hab. West Africa.

a, b. Purchased, June 14, 1866.

393. *Bubo maculosus* (Vieill.). Spotted Eared Owl.
Hab. Africa.

a. Received in exchange, Aug. 16, 1864.
b. Purchased, July 27, 1864.

394. *Bubo poensis,* Fraser. Fraser's Eagle Owl.
Hab. Fernando Po.

a. Purchased, Aug. 13, 1863.

395. *Bubo leucotis* (Temm.). White Eared Owl.
Hab. West Africa; Gambia.

a, b. Purchased, May 19, 1865.

Genus OTUS.

396. *Otus vulgaris* (Linn.).   Long-eared Owl.
Hab. British Islands.

  a. Presented by Robert Widdowson, Esq., Aug. 19, 1865.
  b–d. Presented by William Burnley, Esq., Aug. 13, 1866.

Genus SCOTOPELIA.

397. *Scotopelia peli* (Temm.).   Pel's Owl.
Hab. Gambia.

  a. Purchased, April 30, 1866.

Genus NYCTEA.

398. *Nyctea nivea* (Daud.).   Snowy Owl.
Hab. Europe.

  a. Presented by George Clive, Esq., Oct. 11, 1860.   From Ireland.
  b. Presented by the Hon. A. Gordon, April 30, 1864.   From New Brunswick.

Genus SYRNIUM.

399. *Syrnium aluco* (Linn.).   Wood-Owl.
Hab. Europe.

  a. Presented by Edward Newton, Esq., F.Z.S., May 28, 1851. From Norway.
  b. Presented by Percy S. Godman, Esq., C.M.Z.S., Aug. 13, 1861.   From Norway.
  c. Presented by John Gould, Esq., V.P.Z.S., April 10, 1865.
  d. Presented by the Rev. John Light, Aug. 22, 1866.

400. *Syrnium nebulosum* (Forst.).   Barred Owl.
Hab. North America.

  a. Purchased, Aug. 12, 1863.

401. *Syrnium seloputo* (Horsf.).   Pagoda Owl.
Hab. Java.

  a. Received in exchange, Nov. 10, 1865.

Genus ATHENE.

402. *Athene noctua* (Retz.).   Naked-footed Owl.
Hab. Europe.

a–h. Presented by the Hon. W. C. Ellis, May 23, 1864.
i–j. Purchased, July 9, 1866.

403. *Athene boobook* (Lath.).   Boobook Owl.
Hab. Queensland.

a. Presented by the Acclimatization Society of Queensland,
Sept. 9, 1864.

404. *Athene torquata* (Daud.).   Downy Owl.
Hab. South America.

a. Deposited, Oct. 24, 1864.
b. Purchased, April 8. 1865.

Genus STRIX.

405. *Strix flammea*, Linn.   Common Barn-Owl.
Hab. British Isles.

a. Presented by Mrs. Smale, Sept. 19, 1856.
b. Purchased, Jan. 1, 1863.

406. *Strix pratincola*, Bonap.   American Barn-Owl.
Hab. North America.

a, b, c. Presented by Dr. Slack, of Philadelphia, Jan. 1, 1863.

407. *Strix capensis*, A. Smith.   Cape Barn-Owl.
Hab. South Africa.

a–e. Presented by W. C. Bird, Esq., Nov. 23, 1865.

Order COLUMBÆ.

Family COLUMBIDÆ.

Genus CARPOPHAGA.

408. *Carpophaga globicera*, Wagler.   Wattled Fruit-Pigeon.
Hab. Samoan Islands.

   a. Purchased, Oct. 20, 1862.
   b, c. Purchased, April 19, 1866.

409. *Carpophaga ænea* (Linn.).   Bronze Fruit-Pigeon.
Hab. India.

   a. Purchased, Jan. 1, 1866.

410. *Carpophaga latrans*, Peale.   Brown-tailed Fruit-Pigeon.
Hab. Society Islands.

   a, b. Purchased, April 19, 1866.

Genus PTILONOPUS.

411. *Ptilonopus superbus*, Temm.   Superb Fruit-Pigeon.
Hab. North Australia.

   a, b. Purchased, July 6, 1865.

412. *Ptilonopus melanocephalus*, Gm.   Black-headed Fruit-
Pigeon.
Hab. Java.

   a, b. Females.   Purchased, Oct. 20, 1865.
   c. Male.   Purchased, July 5, 1866.

Genus ERYTHRŒNAS.

413. *Erythrœnas pulcherrima* (Scop.).   Red-crowned Pigeon.
Hab. Seychelles.

   a, b. Presented by Lady Barkly, March 17, 1865.

Genus TRERON.

414. *Treron bicincta* (Jerd.). Double-banded Pigeon.
Hab. Madras.

　a. Female. Deposited, July 7, 1864.
　b. Presented by the Babu Rajendra Mullick, C.M.Z.S., Dec. 22, 1864.

415. *Treron phœnicoptera* (Lath.). Purple-shouldered Pigeon.
Hab. Nepal.

　a. Deposited, July 7, 1864.
　b. Presented by the Babu Rajendra Mullick, C.M.Z.S., Dec. 22, 1864.

416. *Treron macrorhyncha*, Fraser. Thick-billed Pigeon.
Hab. West Africa.

　a. Purchased, Nov. 27, 1865.

Genus LOPHOLÆMUS.

417. *Lopholæmus antarcticus* (Shaw). Double-crested Pigeon.
Hab. North Australia.

　a. Presented by the Acclimatization Society of Sydney, April 10, 1864.
　b. Presented by the Acclimatization Society of New South Wales, May 5, 1865.

Genus COLUMBA.

418. *Columba œnas*, Linn. Stock-Dove.
Hab. British Islands.

　a, b. Presented by Mr. E. Bartlett, Nov. 11, 1862.
　c. Presented by John Fletcher, Esq., Nov. 9, 1863.
　d. Presented by H. B. Bellamy, Esq., Jan. 9, 1865.

419. *Columba livia*, Linn. Rock-Pigeon.
Hab. British Islands.

　a, b. Presented by W. B. Tegetmeier, Esq., Aug. 3, 1866.

420. *Columba domestica*, Gm., var.   Archangel Pigeon.
Hab. British Islands.

a–o. Deposited, Jan. 16, 1860.
p–u. Presented by the Babu Rajendra Mullick, C.M.Z.S.,
March 17, 1864.

421. *Columba gymnophthalma*, Temm.   Naked-eyed Pigeon.
Hab. South America.

a. Presented by W. D. Christie, Esq., F.Z.S., Oct. 4, 1858.
b. Bred in the Gardens, 1858.
c. Hybrid between this species and *O. maculosa*, Temm.
Bred in the Gardens, Aug. 11, 1859.
d. Bred in the Gardens, 1862.
e. Bred in the Gardens, 1863.

422. *Columba albilineata* (Bonap.).   White-naped Pigeon.
Hab. South America.

a. Purchased, Oct. 19, 1851.

423. *Columba leucocephala*, Linn.   White-crowned Pigeon.
Hab. West Indies.

a. Purchased, May 10, 1865.
b, c. Males ; *d, e.* Females.   Presented by John A. Palin, Esq.,
R.M.S. 'Tasmania,' March 31, 1866.
f. Bred in the Gardens, Aug. 1, 1866.
g–h. Presented by Dr. Huggins, C.M.Z.S., Oct. 15, 1866.

424. *Columba guinea*, Linn.   Triangular-spotted Pigeon.
Hab. West Africa.

a, b. Purchased, May 19, 1865.
c. Male ; *d.* Female.   Purchased, Nov. 10, 1866.

425. *Columba arquatrix*, Temm.   Spotted Pigeon.
Hab. South Africa.

a. Purchased, May 7, 1864.

Genus ECTOPISTES.

426. *Ectopistes migratorius* (Linn.).   Passenger Pigeon.
Hab. North America.

a. Female.   Received in exchange, March 8, 1852.
b. Male.   Received in exchange, July 8, 1857.
c. Female.   Purchased, Aug. 3, 1860.

K

Genus ZENAIDURA.

427. *Zenaidura carolinensis* (Linn.).   Carolina Dove.
Hab. North America.

   a. Purchased, 1861.

Genus TURTUR.

428. *Turtur auritus*, Gray.   Common Turtle Dove.
Hab. British Islands.

   a, b. Deposited, 1863.
   c, d. Hybrids between this species and *Turtur risorius* (Linn.). Purchased, Dec. 15, 1863.
   e–g. Presented by Mrs. Sidney, May 19, 1865.

429. *Turtur meena*, Sykes.   Eastern Turtle Dove.
Hab. India.

   a. Presented by the Babu Rajendra Mullick, C.M.Z.S., March 17, 1864.

430. *Turtur picturatus* (Temm.).   Painted Dove.
Hab. Mauritius.

   a, b. Presented by Mrs. Moon, May 11, 1866.

431. *Turtur senegalensis* (Linn.).   Cambayan Turtle Dove.
Hab. Egypt.

   a. Bred in the Gardens, 1861.
   b, c. Bred in the Gardens, Sept. 1, 1862.
   d, e. Bred in the Gardens, Aug. 10, 1862.
   f. Bred in the Gardens, 1863.
   g. Bred in the Gardens, May 22, 1865.

432. *Turtur vinaceus* (Gm.).   Vinaceous Turtle Dove.
Hab. West Africa.

   a. Bred in the Gardens, Oct. 18, 1858.
   b. Bred in the Gardens, July 15, 1860.
   c. Bred in the Gardens, Sept. 26, 1860.
   d. Bred in the Gardens, June 27, 1861.
   e–l. Bred in the Gardens, 1863.
   m, n. Bred in the Gardens, June 27, 1865.
   o, p. Bred in the Gardens, Sept. 4, 1865.
   q, r. Bred in the Gardens, Nov. 17, 1865.
   s, t. Bred in the Gardens, June 20, 1866.

u. Bred in the Gardens, Aug. 1, 1866.
v. Bred in the Gardens, Sept. 29, 1866.
w, x. Bred in the Gardens, Oct. 17, 1866.

433. *Turtur risorius* (Linn.).   Barbary Turtle Dove.
Hab. Africa and India.

a. Presented by Mrs. Bruce, April 14, 1862.
b. Hybrid.   Presented by the Rev. — Delmar, May 31, 1862.
c–f. Deposited, Nov. 16, 1865.

434. *Turtur bitorquatus* (Temm.).   Double-ringed Turtle
Dove.
Hab. Java.

a. Received in exchange, Jan. 27, 1863.

435. *Turtur humilis* (Temm.).   Dwarf Turtle Dove.
Hab. India.

a, b. Presented by M. Jules Verreaux, C.M.Z.S., July 6, 1862.
c–h. Presented by the Babu Rajendra Mullick, C.M.Z.S.,
March 17, 1864.
i, j. Bred in the Gardens, June 3, 1864.

436. *Turtur,* sp.?
Hab. India.

a–c. Presented by A. Macdonald, Esq., July 20, 1865.

Genus ZENAIDA.

437. *Zenaida amabilis*, Bonap.   Zenaida Dove.
Hab. North America.

a. Purchased, 1861.
b, c. Presented by Capt. Sawyer, R.M.S. 'Tasmania,' March 31,
1866.
d. Male ; *e.* Female.   Presented by John A. Palm, Esq.,
R.M.S. 'Tasmania,' March 31, 1866.

Genus CHAMÆPELIA.

438. *Chamæpelia passerina* (Linn.).   Passerine Ground-Dove.
Hab. North America.

a. Presented by Sir Charles S. Smith, Sept. 29, 1860.
b. Presented by T. Goin, Esq., Jan. 1, 1864.

Genus ŒNA.

439. *Œna capensis* (Linn.). Cape Dove.
Hab. Africa.

a, b. Purchased, May 19, 1865.
c–h. Presented by Henry C. Calvert, Esq. From Djeddah,
Arabia, Sept. 5, 1865.

Genus GEOPELIA.

440. *Geopelia placida,* Gould. Placid Ground-Dove.
Hab. East Indies.

a. Presented by Mrs. Sparkes, Nov. 15, 1864.

441. *Geopelia striata* (Linn.). Barred Dove.
Hab. India.

a, b. Purchased, Oct. 2, 1863.
c–e. Presented by Arthur Dixon, Esq., Feb. 23, 1864. From
Japan.
f–j. Presented by the Babu Rajendra Mullick, C.M.Z.S.,
March 17, 1864.
k. Bred in the Gardens, Aug. 29, 1865.

Genus STARNŒNAS.

442. *Starnœnas cyanocephala* (Linn.). Blue-headed Pigeon.
Hab. West Indies.

a, b. Purchased, July 20, 1864.

Genus GEOTRYGON.

443. *Geotrygon montana* (Linn.). Red Ground-Dove.
Hab. Brazil.

a, b. Purchased, Aug. 23, 1860.
c. Bred in the Gardens, June 13, 1863.
d. Bred in the Gardens, July 18, 1864.
e. Bred in the Gardens, June 2, 1865.

444. *Geotrygon sylvatica,* Gosse. Mountain Witch-Dove.
Hab. Jamaica.

a. Purchased, Aug. 23, 1860.
b, c, d. Purchased, July 30, 1861.

445. *Geotrygon mystacea* (Temm.). Moustache Pigeon.
Hab. Martinique.

a–d. Purchased, July 20, 1864.

Genus OCYPHAPS.

446. *Ocyphaps lophotes* (Temm.). Crested Pigeon.
Hab. Australia.

a. Bred in the Gardens, Aug. 21, 1859.
b. Purchased, March 14, 1863.
c. Purchased, Jan. 19, 1865.
d, e. Purchased, Feb. 7, 1865.
f. Purchased, March 7, 1865.
g, h. Bred in the Gardens, July 4, 1865.
i. Purchased, May 17, 1866.
j. Bred in the Gardens, Aug. 1, 1866.
k, l. Bred in the Gardens, Sept. 29, 1866.

Genus CHALCOPHAPS.

447. *Chalcophaps indica* (Linn.). Green-winged Dove.
Hab. India.

a. Received in exchange. From China.
b, c. Presented by the Babu Rajendra Mullick, C.M.Z.S.,
March 17, 1864.
d, e, f. Received, Dec. 31, 1866.

448. *Chalcophaps chrysochlora* (Wagl.). Little Green-winged
Dove.
Hab. Australia.

a. Purchased, Dec. 11, 1863.
b. Purchased, Feb. 18, 1861.
c–g. Received in exchange, April 23, 1863.

Genus PHAPS.

449. *Phaps chalcoptera* (Lath.). Bronze-winged Pigeon.
Hab. Australia.

a. Presented by George Macleay, Esq., F.Z.S., June 22, 1859.
b. Bred in the Gardens, July 19, 1861.
c. Bred in the Gardens, June 20, 1862.

d, e. Deposited, Feb. 14, 1863.
f. Deposited, Nov. 14, 1863.
g. Male. Presented by Miss Ada Angus, April 9, 1865.
h, i. Bred in the Gardens, May 26, 1865.
j, k. Bred in the Gardens, June 2, 1865.
l, m. Bred in the Gardens, July 4, 1865.
n, o. Bred in the Gardens, Sept. 26, 1865.
p. Bred in the Gardens, May 29, 1866.
q. Bred in the Gardens, July 10, 1866.
r–t. Deposited, July 20, 1866.
u, v. Bred in the Gardens, Oct. 6, 1866.

450. *Phaps histrionica*, Gould.   Harlequin Bronze-winged
    Pigeon.

Hab. South Australia.

a–c. Presented by the Acclimatization Society of Victoria,
    Jan. 4, 1865.
d. Bred in the Gardens, June 16, 1866.

Genus CHALCOPELIA.

451. *Chalcopelia chalcospilos* (Wagl.).   Bronze-spotted Dove.
Hab. West Africa.

a, b. Males; *c, d.* Females. Presented by the Hon. Lady
    Cust, Nov. 28, 1866.
e. Male; *f.* Female. Deposited, Nov. 28, 1866.

Genus LEUCOSARCIA.

452. *Leucosarcia picata* (Lath.).   Wonga-wonga Pigeon.
Hab. New South Wales.

a, b. Presented by George Macleay, Esq., F.Z.S., June 2, 1859.
c. Bred in the Gardens, Aug. 11, 1859.
d. Bred in the Gardens, July 23, 1863.
e–g. Deposited, July 20, 1866.

Genus PHLOGŒNAS.

453. *Phlogœnas cruentata* (Lath.).   Red-breasted Pigeon.
Hab. Philippine Islands.

a. Purchased, Sept. 10, 1863.
b, c. Purchased, Dec. 11, 1863.

454. *Phlogœnas crinigera*, Puch.   Bartlett's Pigeon.
Hab. Sooloo Islands.

a-d. Purchased, Aug. 17, 1863.
e. Bred in the Gardens, July 18, 1864.
f. Bred in the Gardens, Sept. 22, 1864.
g. Bred in the Gardens, May 22, 1865.
h. Bred in the Gardens, June 27, 1865.
i-k. Bred in the Gardens, Aug. 29, 1865.

Genus CALŒNAS.

455. *Calœnas nicobarica* (Linn.).   Nicobar Pigeon.
Hab. Indian Archipelago.

a-f. Presented by the Babu Rajendra Mullick, C.M.Z.S., March 17, 1864.
g. Bred in the Gardens, July 6, 1865.
h. Bred in the Gardens, May 25, 1866.
i. Bred in the Gardens, July 31, 1866.

Genus GOURA.

456. *Goura coronata* (Linn.).   Common Crowned Pigeon.
Hab. New Guinea.

a, b. Presented by the Babu Rajendra Mullick, C.M.Z.S., March 17, 1864.

457. *Goura victoriæ*, Fraser.   Victoria Crowned Pigeon.
Hab. Island of Jobie.

a. Male.   Purchased, Sept. 1, 1856.
b, c. Purchased, Sept. 25, 1861.

Genus DIDUNCULUS.

458. *Didunculus strigirostris*, Jard.   Tooth-billed Pigeon.
Hab. Samoan Islands.

a. Female.   Presented by Dr. G. Bennett, F.Z.S., April 10, 1864.

Order GALLINÆ.

Family PTEROCLIDÆ.

Genus PTEROCLES.

459. *Pterocles alchata* (Linn.).　Pintailed Sand-Grouse.
Hab. South Europe.

a. Male.　Purchased, Oct. 7, 1861.
b. Female.　Deposited, 1861.
c–e. Received in exchange, Jan. 12, 1863.
f. Bred in the Gardens, Aug. 29, 1865.

460. *Pterocles exustus,* Temm.　Lesser Pintailed Sand-Grouse.
Hab. North Africa.

a, b. Males; c, d. Females.　Purchased, Aug. 19, 1865.

461. *Pterocles arenarius* (Pall.).　Common Sand-Grouse.
Hab. Asia.
a. Female.　Purchased, June 27, 1864.

462. *Pterocles bicinctus,* Temm. Double-banded Sand-Grouse.
Hab. Senegal.

a–d. Purchased, Dec. 21, 1864.

463. *Pterocles lichtensteinii,* Temm.　Lichtenstein's Sand-
Grouse.
Hab. Djeddah, Arabia.

a–f. Presented by Henry C. Calvert, Esq., Sept. 5, 1865.

Genus SYRRHAPTES.

464. *Syrrhaptes paradoxus* (Pall.).　Pallas's Sand-Grouse.
Hab. China.

a. Presented by the Hon. J. F. Stuart Wortley, April 15, 1861.
From China.
b. Presented by Alfred Newton, Esq., F.Z.S., June 9, 1863.
From Elveden, Suffolk, England.
c. Presented by Lord Francis Conyngham, Dec. 11, 1863.
From Donegal, Ireland.
d. Presented by Harrington Russell, Esq., M.P., Feb. 25, 1865.

Family TETRAONIDÆ.

Subfamily TETRAONINÆ.

Genus TETRAO.

465. *Tetrao cupido*, Linn. Prairie-Grouse.
Hab. North America.

a–l. Purchased, Dec. 15, 1864.

466. *Tetrao tetrix*, Linn. Black Grouse.
Hab. Europe.

a, b. Males. Presented by John Wingfield Malcolm, Esq., March 13, 1865.
c. Male; d. Female. Purchased, Oct. 19, 1865.

467. *Tetrao urogallus*, Linn. Capercailzie.
Hab. Europe.

a. Purchased, Aug. 29, 1865.

Family PHASIANIDÆ.

Subfamily PERDICINÆ.

Genus GALLOPERDIX.

468. *Galloperdix lunulata* (Valenc.). Hardwicke's Francolin.
Hab. India.

a, b. Males; c–e. Females. Presented by the Babu Rajendra Mullick, C.M.Z.S., March 31, 1863.

469. *Galloperdix spadicea* (Gm.). Rufous Francolin.
Hab. India.

a. Male. Presented by Col. Charles Denison, F.Z.S., Dec. 3, 1863.
b. Female. Purchased, May 7, 1864.

Genus FRANCOLINUS.

470. *Francolinus vulgaris*, Steph. Black Francolin.
Hab. India.

a–e. Presented by the Babu Rajendra Mullick, C.M.Z.S., March 17, 1864.

f, g. Deposited, March 23, 1865.
h. Female. Purchased, May 10, 1866.

471. *Francolinus madagascariensis* (Scop.). Madagascar
Francolin.
Hab. Madagascar.

a–c. Deposited, July 25, 1864.
d–f. Presented by Edward Newton, Esq., C.M.Z.S., March 17,
1865.

472. *Francolinus capensis,* Gm. Cape Francolin.
Hab. South Africa.

a. Presented by H.E. Sir George Grey, K.C.B., F.Z.S., Governor of New Zealand, Oct. 3, 1859.

473. *Francolinus clappertonii* (Childr.). Clapperton's Francolin.
Hab. West Africa.

a–c. Purchased, Nov. 17, 1862.

474. *Francolinus ponticerianus* (Gm.). Grey Francolin.
Hab. India.

a. Purchased, April 5, 1862.

475. *Francolinus gularis* (Shaw). Wood Francolin.
Hab. East Bengal.

a–e. Presented by the Babu Rajendra Mullick, C.M.Z.S.,
March 17, 1864.
f. Purchased, May 7, 1864.
g. Presented by the Babu Rajendra Mullick, C.M.Z.S., Dec. 22,
1864.

Genus ARBORICOLA.

476. *Arboricola torqueola* (Val.). Hill-Francolin.
Hab. North India.

a–c. Deposited, April 4, 1864.
d, e. Deposited, Feb. 1, 1865.
f. Deposited, Sept. 22, 1865.

Genus TETRAOGALLUS.

477. *Tetraogallus himalayensis,* Gray.   Himalayan Snow-
       Partridge.
    Hab. Himalaya Mountains.

    a–d. Deposited, Feb. 1, 1865.
    e. Deposited, March 7, 1865.

Genus PERDIX.

478. *Perdix cinerea* (Linn.).   Common Partridge.
    Hab. British Islands.

    a, b. Presented by — Riley, Esq., Aug. 19, 1864.

Genus CACCABIS.

479. *Caccabis saxatilis,* Bechst.   Greek Partridge.
    Hab. Aleppo.

    a–c. Presented by A. H. Layard, Esq., M.P., Aug. 31, 1863.
    d–g. Purchased, Aug. 28, 1864.
    h. Presented by A. H. Layard, Esq., M.P., Dec. 9, 1864.
        From Alexandretta.
    i–l. Presented by Mrs. Brooks, Nov. 30, 1865.

480. *Caccabis chukar* (Gray).   Chukar Partridge.
    Hab. North-west India.

    a, b. Presented by the Babu Rajendra Mullick, C.M.Z.S.,
        Dec. 22, 1864.

481. *Caccabis rufa* (Linn.).   Red-legged Partridge.
    Hab. Europe.

    a, g. Deposited, May 11, 1865.

482. *Caccabis heyi* (Temm.).   Hey's Partridge.
    Hab. Djeddah, Arabia.

    a–g. Presented by Henry C. Calvert, Esq., Sept. 5, 1865.

Genus SYNŒCUS.

483. *Synœcus australis* (Lath.).   Australian Quail.
    Hab. Australia.

    a–d. Presented by Dr. Mueller, C.M.Z.S., Feb. 27, 1861.
        From Melbourne.
    e–g. Bred in the Gardens, July 14, 1864.

Genus COTURNIX.

484. *Coturnix coromandelica* (Gm.). Rain-Quail.
Hab. India.

a, b. Presented by the Maharajah Duleep Singh, F.Z.S., June 24, 1861.
c–l. Presented by F. J. C. Wildash, Esq., May 3, 1865.

485. *Coturnix communis* (Bonn.). Common Quail.
Hab. British Islands.

a, b. Purchased, July 20, 1862.
c. Presented by Lady Cust, July 20, 1862.
d. Presented by Mrs. Frank T. Buckland, Feb. 9, 1865.
e, f. Males; *g, h*. Females. Presented by F. Moreau Hayward, Esq., July 21, 1865.

486. *Coturnix pectoralis*, Gould. Pectoral Quail.
Hab. Australia.

a. Presented by Dr. Mueller, C.M.Z.S., Jan. 27, 1863. From Melbourne.

Subfamily ODONTOPHORINÆ.

Genus ORTYX.

487. *Ortyx virginianus* (Linn.). Virginian Colin.
Hab. North America.

a, b. Presented by A. Downs, Esq., C.M.Z.S., March 5, 1861. From Halifax.
c, d. Males; *e, f*. Females. Deposited, July 25, 1866.

Genus CALLIPEPLA.

488. *Callipepla picta* (Douglas). Plumed Colin.
Hab. California.

a. Female. Purchased, April 25, 1863.

Genus EUPSYCHORTYX.

489. *Eupsychortyx cristatus* (Linn.). Crested Colin.
Hab. Mexico.

a–f. Received, Aug. 9, 1865.
g–j. Purchased, Nov. 10, 1866.

Subfamily PHASIANINÆ.

Genus LOPHOPHORUS.

490. *Lophophorus impeyanus* (Lath.). Impeyan Pheasant.
Hab. Himalaya Mountains.

a, b. Females. Bred in the Gardens, June 10, 1861.
c. Male. Deposited, July 9, 1862.
d, e. Bred in the Gardens, June 24, 1862.
f, g. Males. Deposited, March 31, 1863.
h–k. Bred in the Gardens, 1863.
l–s. Deposited, Feb, 1, 1865.
t. Male ; *u.* Female. Deposited, Feb. 21, 1865.
v, w. Deposited, March 7, 1865.
x. Deposited, March 23, 1865.
y, z. Bred in the Gardens, May 30, 1865.
aa, bb. Bred in the Gardens, June 27, 1865.
cc, dd. Bred in the Gardens, July 16, 1865.
ee, ff. Bred in the Gardens, June 11, 1866.
gg, hh. Bred in the Gardens, July 16, 1866.
ii. Female. Deposited, Oct. 2, 1866.

Genus CROSSOPTILON.

491. *Crossoptilon auritum* (Pall.). Pallas's Eared Pheasant.
Hab. Northern China.

a, b. Males. Presented by Dudley E. Saurin, Esq., July 13, 1866.
c, d. Females. Purchased, Nov. 10, 1866.

Genus PUCRASIA.

492. *Pucrasia macrolopha* (Less.). Pucras Pheasant.
Hab. Simla.

a. Female. Deposited, Feb. 1, 1865.

Genus PHASIANUS.

493. *Phasianus colchicus*, Linn. Common Pheasant.
Hab. British Islands.

a–g. Deposited, 1863.
h. Male ; *i.* Female. (Bohemian variety.) Deposited, 1863.
j. Male. Hybrid between this species and *Gallus domesticus*,
    Linn. Presented by Lord Wharncliffe, F.Z.S., June 4,
    1862.

k–p. Females. Purchased, Jan. 7, 1865.
q–v. Purchased, Feb. 9, 1865.

494. *Phasianus torquatus,* Gm.   Ring-necked Pheasant.
Hab. China.

a. Male. Deposited, June 3, 1864.
b. c. Males ; *d, e.* Females. Deposited, Feb. 17, 1865.
f. Purchased, April 3, 1865.
g. Deposited, April 20, 1865.

495. *Phasianus versicolor,* Vieill.   Japanese Pheasant.
Hab. Japan.

a. Male. Purchased, Sept. 22, 1862.
b. Male ; *c.* Female. Received in exchange, Dec. 18, 1862.
d. Female. Hybrid between this species and *Phasianus colchicus,* Linn. Deposited, Nov. 19, 1862.
e–g. Bred in the Gardens, 1863.
h. Bred in the Gardens, 1864.
i. Female. Deposited, Feb. 17, 1865.
j. Male. Deposited, May 29, 1865.
k–v. Bred in the Gardens, June 27, 1865.

496. *Phasianus sœmmeringii,* Temm.   Sœmmering's Pheasant.
Hab. Japan.

a. Female. Deposited, Dec. 18, 1863.
b, c. Males ; *d, e.* Females. Purchased, June 22, 1864.
f, g. Males. Presented by Reginald Russell, Esq., June 22, 1864.
h. Female. Deposited, May 29, 1865.
i. Bred in the Gardens, June 14, 1865.

497. *Phasianus reevesii,* Gray.   Barred-tailed Pheasant.
Hab. North China.

a. Male. Presented by John Kelk, Esq., F.Z.S., Dec. 8, 1864.
b–d. Males ; *e–h.* Females. Deposited, Aug. 3, 1866.

498. *Phasianus wallichii,* Hardw.   Cheer Pheasant.
Hab. North India.

a. Male. Bred in the Gardens, Aug. 11, 1862.
b. Female. Received in exchange, May 19, 1862.
c. Deposited, March 31, 1863.
d. Bred in the Gardens, July 14, 1863.
e–h. Deposited, Feb. 1, 1865.
i–l. Deposited, March 7, 1865.

m–q. Bred in the Gardens, June 27, 1865.
r–w. Bred in the Gardens, June 11, 1866.

Genus THAUMALEA.

499. *Thaumalea picta* (Linn.). Gold Pheasant.
Hab. China.

Female. Purchased, Feb. 12, 1863.
b. Male; *c.* Female. Purchased, March 18, 1864.
d. Hybrid between this species and *Phasianus colchicus*, Linn.
Deposited, March 30, 1864.
e. Male; *f.* Female. (New variety.) Purchased, July 5,
1865.

Genus EUPLOCAMUS.

500. *Euplocamus prælatus* (Bonap.). Siamese Pheasant.
Hab. Siam.

a. Male. Received in exchange, March 4, 1865.
b–d. Hybrids between this species and *Euplocamus lineatus*
(Lath. MS.). Bred in the Gardens, June 14, 1865.
e. Male; *f.* Female. Deposited, Sept. 22, 1865.
g. Male; *h.* Female. Purchased, Nov. 10, 1866.

501. *Euplocamus swinhoii*, Gould. Swinhoe's Pheasant.
Hab. Formosa.

a. Male. Purchased, June 7, 1865.
b. Male. Purchased, Feb. 10, 1866.
c. Male. Deposited, June 6, 1866.
d. Male. Purchased, Aug. 7, 1866.
e–g. Males; *h.* Female. Deposited, Aug. 23, 1866.
i. Male; *j, k.* Females. Deposited, Oct. 27, 1866.
l, m. Females. Purchased, Nov. 2, 1866.

502. *Euplocamus erythrophthalmus* (Raffl.). Rufous-tailed
Pheasant.
Hab. Malacca.

a, b. Females. Presented by the Babu Rajendra Mullick,
C.M.Z.S., July 25, 1864.

503. *Euplocamus nycthemerus* (Linn.). Silver Pheasant.
Hab. China.

a. Male; *b.* Female. Received in exchange, 1861.
c. Bred in the Gardens, June 3, 1862.
d–f. Deposited, Feb. 17, 1865.

g. Deposited, Sept. 22, 1865.   From India.
h. Female. Hybrid? Deposited, Sept. 22, 1865. From India.
i. Male. Hybrid? Purchased, Jan. 6, 1866.

504. *Euplocamus lineatus* (Lath. MS.).   Lineated Pheasant.
Hab. Tenasserim and Pegu.

a. Male ; *b, c.* Females.   Presented by John Squire, Esq.,
C.M.Z.S., July 25, 1864.
d. Male.   Presented by the Babu Rajendra Mullick, C.M.Z.S.,
Dec. 22, 1864.
e, f. Males ; *g.* Female.   Hybrids.   Presented by the Babu
Rajendra Mullick, C.M.Z.S., Dec. 22, 1864.
h–n. Bred in the Gardens, May 25, 1865.
o–r. Bred in the Gardens, June 27, 1865.
s–w. Bred in the Gardens, June 5, 1866.

505. *Euplocamus horsfieldii,* Gray.   Purple Kaleege.
Hab. North-west Himalayas.

a. Male. Bred in the Gardens, July 9, 1861.
b. Female. Presented by the Babu Rajendra Mullick,C.M.Z.S.,
July 14, 1857.
c–e. Bred in the Gardens, 1862.
f–j. Bred in the Gardens, June 14, 1865.
k–m. Bred in the Gardens, June 5, 1866.

506. *Euplocamus melanotus* (Blyth).   Black-backed Kaleege.
Hab. Sikim.

a. Male.   Presented by the late Viscount Canning, July 14,
1857.
b. Female. Bred in the Gardens, 1858.
c. Bred in the Gardens, 1862.
d–f. Bred in the Gardens, June 14, 1865.
g–k. Bred in the Gardens, June 25, 1866.

507. *Euplocamus albo-cristatus* (Vig.).   White - crested
Kaleege.
Hab. North-west Himalayas.

a. Female.   Presented by the late Viscount Canning, July 14,
1857.
b. Female.   Bred in the Gardens, 1859.
c. Male.   Bred in the Gardens, July 9, 1860.
d. Male.   Bred in the Gardens, July 9, 1861.
e–g. Deposited, March 31, 1863.
h, i. Males ; *j, k.* Females.   Deposited, March 23, 1865.
l–p. Bred in the Gardens, June 14, 1865.
q–u. Bred in the Gardens, June 25, 1866.

Genus GALLUS.

508. *Gallus sonneratii*, Temm.   Sonnerat's Jungle-fowl.
   Hab. India.

   a. Male.   Purchased, July 10, 1860.
   b. Male.   Three-quarters bred.   Bred in the Gardens, Aug. 11, 1862.
   c. Female.   Purchased, March 12, 1863.
   d. Female.   Presented by Col. Charles Denison, F.Z.S., May 29, 1863.
   e–i. Half-bred.   Bred in the Gardens, 1863.
   j. Male.   Received in exchange, March 14, 1865.
   k–r. Bred in the Gardens, May 25, 1865.
   s–w. Bred in the Gardens, June 14, 1865.
   x. Female.   Presented by J. C. Parr, Esq., Nov. 7, 1865.

509. *Gallus bankiva*, Temm.   Bankiva Jungle-fowl.
   Hab. India, generally.

   a–c. Males.   Presented by the Babu Rajendra Mullick, C.M.ZS., Dec. 22, 1864.
   d–i. Females.   Hybrids.   Presented by the Babu Rajendra Mullick, C.M.Z.S., Dec. 22, 1864.
   j–x. Bred in the Gardens, May 25, 1865.
   y. Bred in the Gardens, June 14, 1865.

510. *Gallus domesticus*, Linn., var.   Japanese Fowl.
   Hab. Japan.

   a, b. Males; c. Female.   Bred in the Gardens, 1862.
   d, e. From Africa.   Presented by John Petherick, Esq., C.M.Z.S., June 13, 1866.

Genus CERIORNIS.

511. *Ceriornis satyra* (Linn.).   Horned Tragopan.
   Hab. Himalaya Mountains.

   a. Male; b, c. Females.   Presented by the Babu Rajendra Mullick, C.M.Z.S., March 31, 1863.
   d–g. Males; h. Female.   Deposited, March 31, 1863.
   i, k. Bred in the Gardens, July 7, 1863.
   l. Bred in the Gardens, June 10, 1864.
   m, n. Deposited, March 23, 1865.
   o. Bred in the Gardens, June 27, 1865.
   p–r. Bred in the Gardens, July 16, 1865.
   s. Female.   Purchased, Nov. 11, 1865.

L

512. *Ceriornis temminckii* (Gray). Temminck's Tragopan.
Hab. China.

a. Male. Purchased, Nov. 10, 1864.
b, c. Males. Deposited, Aug. 3, 1866.

Subfamily PAVONINÆ.

Genus PAVO.

513. *Pavo cristatus*, Linn. Common Peafowl.
Hab. India.

a. Male; b. Female. Deposited, Jan. 16, 1860.
c. Male; d. Female. (White variety.) Deposited, Jan. 26, 1860.
e. Presented by A. Grote, Esq., C.M.Z.S., April 30, 1863. From Burmah.
f, g. Females. Hybrids between this species and *Pavo nigripennis*, Sclater. Bred in the Gardens, July 24, 1863.
h, i. (White variety.) Deposited, April 27, 1865.
j. Male. Deposited, Feb. 21, 1866.

514. *Pavo nigripennis*, Sclater. Black-winged Peafowl.
Hab. Sumatra (?).

a, b. Males; c. Female. Deposited, Jan. 16, 1860.
d. Male. Presented by W. R. W. Halsey, Esq., April 29, 1859.

515. *Pavo muticus*, Horsf. Javan Peafowl.
Hab. Burmah.

a. Male. Presented by William Dunn, Esq., C.M.Z.S., July 25, 1864.
b, c. Females. Presented by the Babu Rajendra Mullick, C.M.Z.S., July 25, 1864.

Genus POLYPLECTRON.

516. *Polyplectron chinquis*, Temm. Peacock Pheasant.
Hab. Burmah.

a, b. Males. Presented by the Babu Rajendra Mullick, C.M.Z.S., July 14, 1857.
c. Male. Presented by the Babu Rajendra Mullick, C.M.Z.S., March 31, 1863.
d. Female. Presented by the Babu Rajendra Mullick, C.M.Z.S., July 25, 1864.
e. Bred in the Gardens, May 27, 1866.

Subfamily MELEAGRINÆ.

Genus MELEAGRIS.

517. *Meleagris ocellata*, Temm.   Ocellated Turkey.
  Hab. Guatemala.

   a. Female.   Presented by Robert Owen, Esq., C.M.Z.S.,
      April 29, 1861.
   b. Female.   Presented by Capt. d'Arcy, R.N., Nov. 22, 1864.
   c. Male. Hybrid between this species and *Meleagris mexi-
      cana*, Gould, var. *domestica*. Bred in the Gardens, June 14,
      1865.

518. *Meleagris mexicana*, Gould, var. *domestica*.   Common
      Turkey.
  Hab. Mexico.

   a. Male.   Purchased, May 20, 1862.
   b, c. Presented by O. Miller, Esq., Jan. 20, 1865.

519. *Meleagris gallo-pavo*, Linn.   North-American Turkey.
  Hab. North America.

   a, b. Purchased, Nov. 20, 1863.
   c, d. Presented by E. K. Karslake, Esq., F.Z.S., Nov. 8, 1866.
      From Canada.

Subfamily NUMIDINÆ.

Genus NUMIDA.

520. *Numida meleagris* (Linn.).   Common Guinea-fowl.
  Hab. British Islands.

   a. Presented by Lady Walker, Aug. 26, 1864.   From the Cape
      of Good Hope.
   b, c. White variety.   Presented by Lieut. L. C. Keppel, R.N.,
      Feb. 23, 1864.   From Madagascar.

521. *Numida mitrata*, Pall.   Mitred Guinea-fowl.
  Hab. South Africa.

   a, b. Presented by Lady Walker, Aug. 26, 1864.

522. *Numida tiarata*, Bonap.   Tiara'd Guinea-fowl.
  Hab. Madagascar.

   a–d. Presented by Mrs. Moon, May 11, 1866.

523. *Numida ptilorhyncha*, Licht.   Abyssinian Guinea-fowl.
    Hab. Abyssinia.

    a. Male ; b. Female.   Purchased, March 16, 1865.

524. *Numida cristata*, Pall.   Crested Guinea-fowl.
    Hab. West Africa.

    a. Purchased, Sept. 7, 1865.
    b. Purchased, Sept. 9, 1865.
    c. Purchased, Dec. 28, 1865.
    d. Deposited, Dec. 28, 1865.
    e. Purchased, June 13, 1866.

Family CRACIDÆ.

Subfamily PENELOPINÆ.

Genus PENELOPE.

525. *Penelope purpurascens*, Wagl.   Mexican Guan.
    Hab. Central America.

    a, b. Presented by S. Sandbach Parker, Esq., Oct. 23, 1862.

526. *Penelope pipile*, Jacq.   White-crested Guan.
    Hab. Tropical America.

    a. Presented by the Prince de Joinville, Oct. 13, 1863.
    b. Presented by the Comte and Comtesse d'Eu, Feb. 13, 1865.
    c, d. Presented by Robert Barker, Esq., May 30, 1866. From
        Porto Alegre, Rio Grande, Brazil.

527. *Penelope pileata*, Licht.   Red-breasted Guan.
    Hab. Para.

    a, b. Purchased, March 8, 1864.

528. *Penelope cristata*, Gm.   Crested Guan.
    Hab. Central America.

    a. Presented by Capt. d'Arcy, R.N., Nov. 22, 1864.

529. *Penelope greeyi*, Gray.   Greey's Guan.
    Hab. St. Martha, New Granada.

    a. Purchased, July 14, 1865.   Specimen described and figured,
        P. Z. S. 1866. p. 206, pl. xxii.

Genus ORTALIDA.

530. *Ortalida katraca* (Bodd.). Little Guan.
Hab. South America.

a. Purchased, May 28, 1864.
b–e. Presented by Capt. Sawyer, June 12, 1866.

Subfamily CRACINÆ.

Genus CRAX.

531. *Crax carunculata*, Temm. Yarrell's Curassow.
Hab. South America.

a. Male. Purchased, April 5, 1859.
b. Female. Purchased, March 12, 1861.
c, d. Purchased, May 3, 1865.

532. *Crax globicera*, Linn. Globose Curassow.
Hab. Honduras.

a, b. Females. Presented by H.E. R. W. Keate, Governor of Trinidad, Aug. 9, 1862.
c. Male; d. Female. Presented by R. S. Newall, Esq., Aug. 12, 1864.
e. Presented by Capt. Abbott, Aug. 31, 1864.
f. Male; g. Female. Purchased, Nov. 16, 1865.
h. Presented by Commander Glynn, R.N., Aug. 20, 1866.
i. Male; j. Female. Deposited, Oct. 20, 1866.

533. *Crax alector*, Linn. Crested Curassow.
Hab. Guiana.

a, b. Presented by H.E. R. W. Keate, Governor of Trinidad, Aug. 9, 1862.
c. Presented by W. Duncan Stewart, Esq., June 26, 1861.
d. Purchased, May 3, 1865.
e. Presented by Mrs. Beaumont, April 10, 1866.

534. *Crax fasciolata*, Spix. Banded Curassow.
Hab. South America.

a. Received in exchange, March 12, 1861.
b. Presented by the Prince de Joinville, Oct. 13, 1863.

535. *Crax blumenbachii*, Spix. Blumenbach's Curassow.
Hab. Brazil.

a. Received in exchange, March 12, 1861.

Genus PAUXI.

536. *Pauxi mitu* (Linn.).   Razor-billed Curassow.
Hab. Tropical America.

a. Purchased, Aug. 6, 1860.
b, c. Presented by the Prince de Joinville, Oct. 13, 1863.
d, e. Presented by Sir William Clay, Bart., F.Z.S., Dec. 17,
    1863.

537. *Pauxi tomentosa* (Spix).   Lesser Razor-billed Curassow.
Hab. Brazil.

a, b. Purchased, Jan. 14, 1862.

Family MEGAPODIIDÆ.

Genus TALEGALLA.

538. *Talegalla lathami,* Gray.   Brush-Turkey.
Hab. Australia.

a. Female.   Bred in the Gardens, July 19, 1854.
b. Female.   Bred in the Gardens, Aug. 29, 1860.
c. Female.   Received, April 24, 1863.
d. Presented by C. Moore, Esq., Sept. 22, 1863.
e. Male ; f. Female.   Deposited, April 19, 1866.
g. Male.   Deposited, June 6, 1866.
h. Bred in the Gardens, Sept. 5, 1866.
i. Bred in the Gardens, Sept. 7, 1866.
j. Bred in the Gardens, Sept. 11, 1866.
k. Bred in the Gardens, Sept. 30, 1866.

Genus LEIPOA.

539. *Leipoa ocellata,* Gould.   Mallee Bird.
Hab. Australia.

a. Purchased, Sept. 7, 1865.
b. Purchased, April 4, 1866.
c, d. Deposited, July 20, 1866.

Family TINAMIDÆ.

Genus TINAMUS.

540. *Tinamus variegatus,* Gm.   Variegated Tinamou.
Hab. Brazil.

a. Presented by John Blount, Esq., April 8, 1861.

Genus RHYNCHOTUS.

541. *Rhynchotus rufescens* (Temm.).   Rufous Tinamou.
Hab. Brazil.

a, b. Purchased, June 27, 1864.
c. Presented by the Comte and Comtesse d'Eu, Feb. 13, 1865.
d, e. Presented by Edward Collier, Esq., July 10, 1866.

Order STRUTHIONES.

Family STRUTHIONIDÆ.

Genus STRUTHIO.

542. *Struthio camelus*, Linn.   Ostrich.
Hab. North Africa.

a. Male.   Presented by Her Majesty the Queen, April 4, 1850.
From Morocco.
b. Female.   Presented by Her Majesty the Queen, Feb. 8,
1859.   From Morocco.
c, d. Young.   Presented by E. Herslet, Esq., Oct. 10, 1863.
e. Male.   Purchased, Aug. 19, 1865.

Genus RHEA.

543. *Rhea macrorhyncha*, Sclater.   Great-billed Rhea.
Hab. South America.

a. Male.   Purchased, Nov. 1858.   Specimen described and
figured, P. Z. S., 1860, p. 207, and Trans. Z. S. iv. p. 356,
pl. lxix.

544. *Rhea americana*, Vieill.   Common Rhea.
Hab. South America.

a. Male; *b.* Female.   Purchased, Dec. 5, 1860.
c. Young.   Hybrid between this species and *Rhea macrorhyncha*, Sclater.   Bred in the Gardens, 1862.
d. Bred in the Gardens, Aug. 14, 1863.
e, f. Presented by A. Brenner, Esq., July 18, 1864.

Family CASUARIIDÆ.

Genus CASUARIUS.

545. *Casuarius galeatus* (Vieill.).   Common Cassowary.
Hab. Ceram.

a. Male.   Received in exchange, March 28, 1862.

b. Female. Purchased, Nov. 20, 1863.
c. Male; *d.* Female. Presented by the Babu Rajendra Mul-
    lick, C.M.Z.S., July 25, 1864.
e. Bred in the Gardens, June 22, 1866.

546. *Casuarius bennettii*, Gould.   Bennett's Cassowary, or
    Mooruk.

Hab. New Britain.

a. Male.   Presented by Dr. George Bennett, F.Z.S., May 17,
    1857.   Specimen figured, Trans. Z. S. iv. pl. lxxii.
b. Female.   Presented by Dr. George Bennett, F.Z.S., May 25,
    1858.
c, d. Bred in the Gardens, June 21, 1864.

Genus DROMÆUS.

547. *Dromæus novæ-hollandiæ*, Vieill.   Emu.
Hab. New South Wales.

a. Male.   Received in exchange, April 1862.
b. Presented by D. P. McEwen, Esq., June 14, 1864.
c, d. Presented by H. G. Ashurst, Esq., Aug. 3, 1864.
e, f. Var. *irrorata*, Bartlett.   Bred in the Gardens, 1861.
g–j. Bred in the Gardens, March 31, 1865.
k–m. Presented by Dr. Mueller, C.M.Z.S., July 26, 1865.
n. Presented by the Hon. Mrs. Greville Howard, Oct. 6, 1865.

Family APTERYGIDÆ.

Genus APTERYX.

548. *Apteryx mantellii*, Bartl.   Kiwi.
Hab. New Zealand.

a. Female.   Presented by Lieut. Governor Eyre, Dec. 9, 1851.
b. Male.   Presented by Major Keane, Sept. 29, 1864.
c. Presented by Henry Slade, Esq., R.N., May 23, 1865.

Order GRALLÆ.

Family OTIDÆ.

Genus OTIS.

549. *Otis tarda*, Linn.   Great Bustard.
Hab. Europe.

>*a, b.* Males.   Purchased, Oct. 10, 1863.
>*c.* Female.   Purchased, Nov. 7, 1863.

550. *Otis australis* (Gray).   Australian Bustard.
Hab. Australia.

>*a.* Purchased, April 4, 1866.
>*b.* Presented by the Acclimatization Society of Sydney, April 19, 1866.

551. *Otis kori*, Burch.   Burchell's Bustard.
Hab. South Africa.

>*a.* Purchased, May 10, 1866.

Genus TETRAX.

552. *Tetrax campestris*, Leach.   Little Bustard.
Hab. Europe.

>*a.* Received in exchange, Sept. 20, 1861.
>*b.* Received in exchange, Oct. 24, 1862.
>*c, d.* Purchased, 1863.
>*e, f.* Deposited, Oct. 29, 1866.

Genus EUPODOTIS.

553. *Eupodotis bengalensis* (Gm.).   Bengal Bustard.
Hab. Bengal.

>*a.* Male.   Presented by the Babu Rajendra Mullick, C.M.Z.S., July 14, 1857.

Family CHARADRIIDÆ.

Genus VANELLUS.

554. *Vanellus cristatus,* Meyer.   Peewit.
Hab. British Islands.

    a. Presented by — Bergmann, Esq., Aug. 26, 1863.
    b. Presented by A. Yates, Esq., Aug. 2, 1864.
    c. Presented by F. Cresswell, Esq., Dec. 5, 1866.

Genus SARCIOPHORUS.

555. *Sarciophorus pectoralis* (Cuv.).   Black-breasted Peewit.
Hab. Australia.

    a, b. Deposited, May 23, 1864.

Genus LOBIVANELLUS.

556. *Lobivanellus lobatus* (Lath.).   Wattled Peewit.
Hab. Australia.

    a, b. Purchased, April 19, 1866.

Genus ŒDICNEMUS.

557. *Œdicnemus crepitans,* Temm.   Thicknee.
Hab. British Islands.

    a. Presented by Lord Lilford, F.Z.S., June 10, 1862.
    b. Presented by R. M. Presland, Esq., Aug. 2, 1865.

558. *Œdicnemus grallarius* (Lath.).   Australian Thicknee.
Hab. Australia.

    a, b. Presented by the Royal Dublin Zoological Society,
        April 10, 1864.
    c. Deposited, July 20, 1866.

Genus CHARADRIUS.

559. *Charadrius pluvialis,* Linn.   Golden Plover.
Hab. Europe.

    a. Presented by F. Cresswell, Esq., Nov. 30, 1864.
    b. Deposited, March 15, 1865.
    c. Presented by F. Cresswell, Esq., Dec. 5, 1866.

Genus SQUATAROLA.

560. *Squatarola cinerea* (Ray). Grey Plover.
Hab. British Islands.

    a. Deposited, April 15, 1865.
    b. Deposited, April 30, 1866. From Norfolk.

Genus HÆMATOPUS.

561. *Hæmatopus ostralegus*, Linn. Oyster-catcher.
Hab. Europe, Asia, Africa.

    a–d. Purchased, March 27, 1862.
    e, f. Received in exchange, March 10, 1865.
    g. Deposited, April 30, 1866. From Norfolk.

Family CHIONIDIDÆ.

Genus CHIONIS.

562. *Chionis alba*, Forst. Sheathbill.
Hab. Antarctic America.

    a. Purchased, July 4, 1865.

Family SCOLOPACIDÆ.

Genus NUMENIUS.

563. *Numenius arquatus*, Linn. Common Curlew.
Hab. Europe, Asia, Africa.

    a. Purchased, March 6, 1860.
    b–e. Received in exchange, March 10, 1865.
    f. Received in exchange, April 5, 1866.

564. *Numenius minutus*, Gould. Little Whimbrel.
Hab. Australia.

    a. Purchased, April 19, 1866.

Genus LIMOSA.

565. *Limosa lapponica* (Linn.). Bar-tailed Godwit.
Hab. Europe.

    a. Deposited, Aug. 11, 1860.
    b. Deposited, March 15, 1865.

566. *Limosa melanura*, Leisler.   Black-tailed Godwit.
Hab. Europe.

  a. Purchased, Aug. 10, 1860.

Genus MACHETES.

567. *Machetes pugnax* (Linn.).   Ruff.
Hab. Europe.

  a–e. Purchased, May 6, 1862.

Genus TRINGA.

568. *Tringa variabilis*, Meyer.   Dunlin.
Hab. Europe.

  a–o. Presented by F. Cresswell, Esq., Nov. 30, 1864.
  p–s. Presented by F. Cresswell, Esq., Dec. 5, 1866.

Genus CALIDRIS.

569. *Calidris canutus*, Briss.   Knot.
Hab. Europe.

  a–c. Presented by F. Cresswell, Esq., Nov. 30, 1864.
  d–l. Presented by F. Cresswell, Esq., Dec. 5, 1866.

Genus TOTANUS.

570. *Totanus hypoleucos* (Linn.).   Common Sandpiper.
Hab. British Islands.

  a. Presented by R. Mitford, Esq., June 7, 1865.

Genus STREPSILAS.

571. *Strepsilas interpres* (Linn.).   Turnstone.
Hab. Europe.

  a. Purchased, Sept. 7, 1865.
  b. Presented by F. Cresswell, Esq., Dec. 5, 1866.

Family PSOPHIIDÆ.

Genus PSOPHIA.

572. *Psophia crepitans*, Linn.   Common Trumpeter.
Hab. Demerara.

  a. Purchased, Jan. 14, 1864.

b–d. Presented by J. Lucie Smith, Esq., July 5, 1866.
e. Purchased, July 13, 1866. From Surinam.

573. *Psophia viridis,* Spix. Green-winged Trumpeter.
Hab. Brazil.

    a, b. Purchased, March 8, 1864.
    c. Purchased, May 10, 1866.

Family CARIAMIDÆ.

Genus CARIAMA.

574. *Cariama cristata,* Linn. Cariama.
Hab. Tropical America.

    a. Presented by William Downing, Esq., Aug 10, 1863.
    b. Purchased, Oct. 10, 1866.

Family GRUIDÆ.

Genus GRUS.

575. *Grus montignesia,* Bp. Mantchurian Crane.
Hab. North China.

    a. Female. Received in exchange, Oct. 20, 1856.

576. *Grus antigone,* Linn. Sarus Crane.
Hab. North India.

    a. Presented by the Babu Rajendra Mullick, C.M.Z.S., July 25, 1864.

577. *Grus australasiana,* Gould. Australian Crane.
Hab. Australia.

    a. Male. Presented by the Marchioness of Londonderry, Feb. 14, 1857.
    b. Female. Received in exchange, April 18, 1863.
    c, d. Presented by the Acclimatization Society of Victoria, June 27, 1864.
    e–g. Deposited, July 20, 1866.

578. *Grus cinerea,* Bechst. Common Crane.
Hab. Europe.

    a. Female. Purchased, May 13, 1848.
    b. Male. Received in exchange, May 27, 1852.
    c. Bred in the Gardens, June 23, 1863.

579. *Grus canadensis* (Linn.).   Brown Crane.
Hab. North America.

a, b. Purchased, Aug. 3, 1866.

580. *Grus carunculata* (Gm.).   Wattled Crane.
Hab. South Africa.

a, b. Received in exchange, Oct. 15, 1866.

Genus TETRAPTERYX.

581. *Tetrapteryx paradiseus,* Licht.   Stanley Crane.
Hab. South Africa.

a. Presented by H.E. Sir George Grey, K.C.B., F.Z.S.,
Governor of New Zealand, May 26, 1861.
b. Presented by H.E. Sir George Grey, K.C B., F.Z.S.,
Governor of New Zealand, Nov. 1, 1861.

Genus BALEARICA.

582. *Balearica pavonina,* Briss.   Balearic Crowned Crane.
Hab. North and West Africa.

a. Presented by Viscount Hill, F.Z.S., Feb. 29, 1860.
b. Purchased, Oct. 30, 1866.

583. *Balearica regulorum,* Licht.   Cape Crowned Crane.
Hab. South Africa.

a. Deposited, Feb. 19, 1851.

Genus ANTHROPOIDES.

584. *Anthropoides virgo* (Linn.).   Demoiselle Crane.
Hab. North Africa.

a, b. Purchased, Jan. 1, 1863.

Family EURYPYGIDÆ.

Genus EURYPYGA.

585. *Eurypyga helias* (Pall.).   Sun-bird.
Hab. Trinidad.

a. Presented by Dr. Huggins, C.M.Z.S., Sept. 16, 1861.

b, c. Purchased, Sept. 29, 1862.
d. Bred in the Gardens, July 9, 1865.
e. On approval, Sept. 13, 1865.
f. Bred in the Gardens, Sept. 28, 1865.
g. Presented by James L. Inman, Esq., R.M.S., 'Shannon,'
    April 30, 1866.
h. Bred in the Gardens, May 18, 1866.
i. Bred in the Gardens, July 10, 1866.
j. Bred in the Gardens, Aug, 15, 1866.

Genus RHINOCHETUS.

586. *Rhinochetus jubatus,* Verr. et Des Murs.   Kagu.
    Hab. New Caledonia.

   a. Male.   Presented by Dr. G. Bennett, F.Z.S., April 25, 1862.
   b. Female.   Presented by Dr. G. Bennett, F.Z.S., April 18,
       1863.
   c. Presented by the Acclimatization Society of Sydney,
       April 19, 1866.

Family RALLIDÆ.

Genus RALLUS.

587. *Rallus pectoralis,* Less.   Australian Rail.
    Hab. New Holland.

   a, b. Received in exchange, April 18, 1863.
   c. Purchased, April 19, 1866.

588. *Rallus aquaticus,* Linn.   Water-Rail.
    Hab. British Islands.

   a. Purchased, Nov. 16, 1865.

589. *Rallus* —— ? ——.   Brazilian Rail.
    Hab. Brazil.

   a. Purchased, Nov. 18, 1865.

Genus ARAMIDES.

590. *Aramides cayennensis* (Gm.).   West-Indian Rail.
    Hab. Trinidad.

   a, b. Purchased, May 6, 1864.

Genus CREX.

591. *Crex pratensis*, Bechst.   Land-Rail.
Hab. British Islands.

a. Purchased, July 25, 1864.

Genus PORZANA.

592. *Porzana maruetta* (Leach).   Spotted Crake.
Hab. British Islands.

a. Presented by William Thompson, Esq., Oct. 27, 1866.

Genus OCYDROMUS.

593. *Ocydromus australis* (Sparrm.).   Weka Rail.
Hab. New Zealand.

a. Male.   Purchased, April 29, 1861.
b, c. Presented by the Acclimatization Society of Melbourne,
May 23, 1864.
d-h. Presented by the Acclimatization Society of Victoria,
Jan. 4, 1865.
i, j. Presented by Dr. Mueller, C.M.Z.S., June 30, 1866.
k, l. Deposited, July 20, 1866.

Genus PORPHYRIO.

594. *Porphyrio melanotus*, Temm.   Black-backed Porphyrio.
Hab. Australia.

a, b. Presented by Edward Wilson, Esq., May 31, 1860.
c. Purchased, April 26, 1862.
d-h. Presented by Dr. Mueller, C.M.Z.S., Nov. 25, 1862.
From Melbourne.
i, j. Presented by Henry St. Hill, Esq., Aug. 20, 1864.   From
New Zealand.
k-n. Deposited, April 19, 1866.
o-s. Presented by Dr. Mueller, C.M.Z.S., June 15, 1866.
t, u. Presented by Dr. Mueller, C.M.Z.S., June 30, 1866.

595. *Porphyrio poliocephalus* (Lath.).   Grey-headed Porphyrio.
Hab. Java.

a-d. Presented by the Babu Rajendra Mullick, C.M.Z.S.,
March 17, 1864.
e, f. Purchased, Jan. 1, 1866.

596. *Porphyrio allenii,* Thomp.   Allen's Porphyrio.
     Hab. West Africa.

     a, b. Purchased, Sept. 22, 1863.

597. *Porphyrio hyacinthinus,* Temm.   Hyacinthine Porphyrio.
     Hab. Bagdad.

     a. Purchased, Dec. 18, 1864.

598. *Porphyrio madagascariensis* (Gmel.).   Madagascar Po-
        phyrio.
     Hab. Madagascar.

     a. Purchased, Nov. 10, 1866.

599. *Porphyrio martinicus* (Linn.).   Martinique Waterhen.
     Hab. West Indies.

     a. Purchased, June 18, 1862.
     b. Purchased, June 16, 1865.

Genus TRIBONYX.

600. *Tribonyx ventralis,* Gould.   Black-tailed Water-Hen.
     Hab. Australia.

     a, b. Purchased, April 6, 1864.

Genus GALLINULA.

601. *Gallinula tenebrosa,* Gould.   Sombre Gallinule.
     Hab. Australia.

     a, b. Presented by Dr. Mueller, C.M.Z.S., Jan. 27, 1863.
        From Melbourne.

602. *Gallinula nesiotis,* Sclater.   Island-Hen Gallinule.
     Hab. Tristan d'Acunha.

     a. Presented by H.E. Sir George Grey, K.C.B., F.Z.S.,
        Governor of New Zealand, May 26, 1861.

Genus FULICA.

603. *Fulica cristata,* Lath.   Crested Coot.
     Hab. South Africa.

     a. Female.   Presented by Earl Fitzwilliam, F.Z.S., Feb. 11,
        1858.

M

b–d. Hybrids between this species and *Fulica atra,* Linn. Bred in the Gardens, June 8, 1863.

604. *Fulica atra,* Linn. Common Coot.
Hab. Europe.

a. Male. Purchased, Nov. 15, 1860.
b. Purchased, Jan. 3, 1859.

605. *Fulica australis,* Gould. Australian Coot.
Hab. Australia.

a, b. Presented by Dr. Mueller, C.M.Z.S., Jan. 27, 1863. From Melbourne.

Order HERODIONES.

Family ARDEIDÆ.

Genus ARDEA.

606. *Ardea cinerea,* Linn. Common Heron.
Hab. Europe.

a. Presented by the Hon. C. A. Ellis, F.Z.S., Oct. 15, 1859.
b. Presented by Lord Lilford, F.Z.S., Aug. 1, 1855.
c. Presented by Dr. A. Günther, F.Z.S., June 18, 1865.
d. Presented by W. M. Wyllie, Esq., Aug. 1, 1865.
e. Presented by J. Lane, Esq., Sept. 19, 1865.

607. *Ardea cocoi,* Linn. Cocoi Heron.
Hab. West Indies.

a. Purchased, May 28, 1864.
b. Presented by J. M. Barton, Esq., C.M.Z.S., July 13, 1866.

608. *Ardea goliath,* Temm. Goliath Heron.
Hab. West Africa.

a. Purchased, June 14, 1865.
b. Purchased, June 13, 1866.

Genus EGRETTA.

609. *Egretta candidissima* (Gm.). Snowy Egret.
Hab. Para.

a–c. Purchased, March 8, 1864.

610. *Egretta alba* (Linn.).   Great White Egret.
Hab. Europe.

a, b. Purchased, Nov. 28, 1865.

611. *Egretta leuce* (Ill.).   Great American Egret.
Hab. Para.

a. Purchased, March 8, 1864.

612. *Egretta garzetta* (Linn.).   Little Egret.
Hab. Lower Danube.

a–e. Purchased, Sept. 19, 1864.

Genus BUPHUS.

613. *Buphus comatus* (Pall.).   Squacco Heron.
Hab. Lower Danube.

a–f. Purchased, Sept. 19, 1864.

Genus ARDETTA.

614. *Ardetta minuta* (Linn.).   Little Bittern.
Hab. Newfoundland.   Europe.

a. Purchased, June 8, 1860.
b–e. Presented by Henry Peters, Esq., June 26, 1866.

Genus BUTORIDES.

615. *Butorides virescens* (Linn.).   Green Bittern.
Hab. Para.

a. Purchased, March 8, 1864.

Genus TIGRISOMA.

616. *Tigrisoma brasiliense* (Linn.).   Tiger Bittern.
Hab. West Indies.

a. Purchased, April 28, 1864.

617. *Tigrisoma leucolophum*, Jard.   White-crested Tiger
Bittern.
Hab. West Africa.

a. Purchased, Dec. 11, 1866.

M 2

Genus NYCTICORAX.

618. *Nycticorax europæus*, Steph.   Common Night-Heron.
Hab. Europe.

a–d. Purchased, Aug. 7, 1863.
e. Presented by Henry Peters, Esq., June 19, 1866.

619. *Nycticorax violaceus*, Linn.   Violaceous Night-Heron.
Hab. West Indies.

a. Presented by E. B. Webb, Esq., April 14, 1864.
b, c. Purchased, June 4, 1866.

620. *Nycticorax caledonicus* (Gm.).   Nankeen Night-Heron.
Hab. Australia.

a. Purchased, April 19, 1866.

Family CICONIIDÆ.

Genus CICONIA.

621. *Ciconia nigra*, Ray.   Black Stork.
Hab. Europe.

a. Presented by W. C. Domville, Esq., Aug. 14, 1855.
b. Received in exchange, Oct. 21, 1861.
c, d. Received in exchange, Oct. 14, 1862.

622. *Ciconia alba*, Linn.   White Stork.
Hab. Europe.

a. Purchased, Oct. 31, 1864.
b–e. Purchased, Aug. 25, 1865.
f. Presented by T. M. Hayward, Esq., June 28, 1866.

623. *Ciconia maguari*, Briss.   Maguari Stork.
Hab. South America.

a, b. Purchased, March 8, 1864.   From Para.

624. *Ciconia leucocephala* (Gm.).   White-necked Stork.
Hab. West Africa.

a. Presented by Mrs. D. Campbell, Aug. 10, 1866.

Genus XENORHYNCHUS.

625. *Xenorhynchus australis*, Shaw.   Black-necked Stork.
Hab. Malacca.

a. Male; b. Female.   Presented by A. Grote, Esq., C.M.Z.S.,
July 25, 1864.

626. *Xenorhynchus senegalensis*, Shaw.   Saddle-billed Stork.
Hab. West Africa.

a, b. Purchased, April 17, 1861.

Genus MYCTERIA.

627. *Mycteria americana*, Linn.   American Jabiru.
Hab. Para.

a. Purchased, March 8, 1864.

Genus LEPTOPTILUS.

628. *Leptoptilus crumeniferus* (Cuv.).   Marabou Stork.
Hab. West Africa.

a. Purchased, Aug. 15, 1854.
b. Presented by Edmund Gabriel, Esq., H.B.M.'s Commissioner
at Loando, Angola, Aug. 22, 1860.   From Angola.
c, d. Presented by Capt. Glover, R.N., June 18, 1866.

Family PLATALEIDÆ.

Genus TANTALUS.

629. *Tantalus ibis*, Linn.   African Tantalus.
Hab. West Africa.

a. Purchased, June 8, 1863.
b. Purchased, July 16, 1863.

630. *Tantalus leucocephalus*, Linn.   Indian Tantalus.
Hab. India.

a. Presented by A. Grote, Esq., C.M.Z.S., July 25, 1864.
b. Presented by the Babu Rajendra Mullick, C.M.Z.S., July 25,
1864.

Family IBIDIDÆ.

Genus IBIS.

631. *Ibis rubra*, Linn.   Scarlet Ibis.
Hab. Para.

a–f. Purchased, March 8, 1864.
g. Purchased, Oct. 15, 1866.

632. *Ibis alba*, Linn.   White Ibis.
Hab. West Indies.

a. Purchased, April 28, 1864.
b. Purchased, June 11, 1866.
c, d. Purchased, Aug. 13, 1866.

633. *Ibis falcinellus*, Linn.   Glossy Ibis.
Hab. Europe.

a, b. Purchased, Nov. 3, 1866.

Genus GERONTICUS.

634. *Geronticus æthiopicus* (Lath.).   Sacred Ibis.
Hab. River Gambia.

a, b. Received in exchange, Aug. 9, 1855.
c, d. Purchased, Nov. 18, 1863.   From South Africa.

635. *Geronticus melanocephalus* (Linn.).   Black-headed Ibis.
Hab. Siam.

a. Purchased, Dec. 18, 1864.

636. *Geronticus calvus* (Bodd.).   Bald-headed Ibis.
Hab. South Africa.

a. Presented by H.E. Sir George Grey, K.C.B., F.Z.S., Governor of New Zealand, Oct. 3, 1859.

637. *Geronticus albicollis* (Gm.).   White-necked Ibis.
Hab. Brazil.

a. Purchased, Nov. 1, 1864.

638. *Geronticus spinicollis*, Jameson.   Straw-necked Ibis.
Hab. Australia.

a. Presented by the Acclimatization Society of Sydney, April 19, 1866.

b. Purchased, April 19, 1866.
c. Presented by the Acclimatization Society of Melbourne, July 21, 1866.

Family PHŒNICOPTERIDÆ.

Genus PHŒNICOPTERUS.

639. *Phœnicopterus antiquorum*, Linn.   Flamingo.
Hab. Egypt.

a–e. Presented by Rees Williams, Esq., June 20, 1862.
f. Presented by Charles Joyce, Esq., April 10, 1865.
g–l. Purchased, July 16, 1866.

640. *Phœnicopterus ruber*, Linn.   Ruddy Flamingo.
Hab. North America.

a–c. Purchased, Aug. 3, 1866.

Order ANSERES.

Family PALAMEDEIDÆ.

Genus CHAUNA.

641. *Chauna derbiana*, Gray.   Derbian Screamer.
Hab. Dekke River, New Granada.

a. Presented by Edward Greey, Esq., F.Z.S., July 29, 1863.
b. Purchased, Nov. 16, 1865.   Specimen described and figured, P. Z. S. 1864, p. 74, pl. xi., as *Chauna nigricollis.*
c. Purchased, June 27, 1866.

Family ANATIDÆ.

Subfamily ANSERINÆ.

Genus ANSERANAS.

642. *Anseranas melanoleuca*, Less.   Black and White Goose.
Hab. Australia.

a. Female.   Purchased, June 19, 1855.
b. Presented by Mrs. R. W. Moore, March 3, 1865.
c, d. Presented by Dr. Mueller, C.M.Z.S., May 5, 1865.

Genus PLECTROPTERUS.

643. *Plectropterus gambensis* (Linn.). Spur-winged Goose.
Hab. West Africa.

a. Female. Purchased, June 25, 1857.

Genus CEREOPSIS.

644. *Cereopsis novæ-hollandiæ*, Lath. Cereopsis Goose.
Hab. Australia.

a. Male. Bred in the Gardens, 1853.
b, c. Females. Purchased, April 26, 1861.
d. Male. Received in exchange, April 17, 1863.
e. Presented by the Acclimatization Society of Victoria, Jan. 4, 1865.
f. Male; *g.* Female. Deposited, March 25, 1865.
h. Male. Presented by John Macmickan, Esq., Dec. 6, 1865.

Genus ANSER.

645. *Anser ferus*, Linn. Wild or Grey-Lag Goose.
Hab. British Islands.

a. Male. Purchased, 1855. From India.
b. Female. Purchased, March 24, 1860.

646. *Anser brachyrhynchus*, Baill. Pink-footed Goose.
Hab. British Islands.

a. Male; *b.* Female. Purchased, March 14, 1861.

647. *Anser segetum*, Linn. Bean-Goose.
Hab. Europe.

a. Male; *b.* Female. Purchased, Nov. 1, 1859.

648. *Anser erythropus*, Linn. Little Goose.
Hab. Europe.

a. Female. Purchased, May 18, 1852.
b. Male. Purchased, May 27, 1852.
c. Male. Purchased, April 4, 1858.

649. *Anser albifrons*, Linn. White-fronted Goose.
Hab. Europe.

a. Male; *b.* Female. Purchased, 1861.

650. *Anser indicus*, Gm. Bar-headed Goose.
Hab. India.

  a. Male; b. Female. Received in exchange, Feb. 5, 1852.

651. *Anser cygnoides*, Linn. Chinese Goose.
Hab. China.

  a. Male. Presented by Capt. Cruikshank, April 22, 1860.
  b. Female. Presented by Russell Sturgis, Esq., F.Z.S., Oct. 7,
    1859.

Genus BERNICLA.

652. *Bernicla leucopsis*, Bechst. Bernicle Goose.
Hab. Europe.

  a. Male. Purchased, 1861.
  b, c. Presented by W. C. Hewitson, Esq., F.Z.S., Aug. 15, 1865.
  d, e. Presented by W. C. Hewitson, Esq., F.Z.S., April 17, 1866.

653. *Bernicla canadensis* (Linn.). Canada Goose.
Hab. British Islands.

  a. Male. Presented by Capt. Wishart, H.B.C.S., Oct. 29,
    1861. From Hudson's Bay Territory.
  b. Presented by Capt. Wishart, H.B.C.S., Oct. 29, 1862. From
    Hudson's Bay Territory.

654. *Bernicla hutchinsii*, Rich. Hutchins's Goose.
Hab. Arctic America.

  a. Female. Presented by Capt. David Herd, H.B.C.S.,
    C.M.Z.S., Oct. 10, 1860. From Fort Churchill, Hudson's
    Bay Territory.

655. *Bernicla ruficollis* (Pall.). Red-breasted Goose.
Hab. Europe.

  a. Female. Received in exchange, Aug. 19, 1858.

656. *Bernicla brenta*, Steph. Brant Goose.
Hab. Europe.

  a, b. Males; c, d. Females. Purchased, March 1, 1864.
  e, f. Males; g, h. Females. Purchased, May 22, 1865.
  i–l. Presented by W. C. Hewitson, Esq., F.Z.S., April 17, 1866.

657. *Bernicla jubata* (Lath.).   Maned Goose.
     Hab. Australia.

     a–c. Purchased, Nov. 21, 1864.
     d–g. Presented by the Acclimatization Society of Victoria.
         Jan. 4, 1865.

Genus CHLOËPHAGA.

658. *Chloëphaga magellanica* (Gm.).   Upland Goose.
     Hab. Falkland Islands.

     a. Male; *b.* Female.   Presented by H.E. Capt. Thomas E. L.
         Moore, R.N., C.M.Z.S., Governor of the Falkland Islands,
         May 27, 1857.
     c. Male; *d.* Female.   Presented by H.E. Capt. Thomas E. L.
         Moore, R.N., C.M.Z.S., Governor of the Falkland Islands,
         Sept. 5, 1861.
     e, f. Females.   Bred in the Gardens, May 4, 1863.
     g, h. Bred in the Gardens, April 30, 1865.

659. *Chloëphaga poliocephala*, Sclater.   Ashy-headed Goose.
     Hab. South America.

     a. Male; *b.* Female.   Bred in the Gardens, June 7, 1858.
     c. Male; *d.* Female.   Bred in the Gardens, May 27, 1860.
     e–h. Bred in the Gardens, May 25, 1865.

660. *Chloëphaga rubidiceps*, Sclater.   Ruddy-headed Goose.
     Hab. Falkland Islands.

     a, b. Males; *c, d.* Females.   Purchased, July 5, 1860.
     e–g. Bred in the Gardens, April 30, 1865.
     h, i. Bred in the Gardens, May 8, 1866.
     j–m. Bred in the Gardens, June 5, 1866.

661. *Chloëphaga sandvicensis* (Vig.).   Sandwich-Island Goose.
     Hab. Sandwich Islands.

     a. Male.   Presented by Earl Fitzwilliam, F.Z.S., Feb. 11, 1858.
     b. Female.   Received, April 25, 1863.
     c–g. Bred in the Gardens, April 5, 1864.
     h–j. Deposited, Jan. 19, 1866.

Genus CHENALOPEX.

662. *Chenalopex ægyptiaca* (Linn.).   Egyptian Goose.
     Hab. Africa.

     a. Deposited, Aug. 11, 1863.   From South Africa.
     b. Presented by W. C. Hewitson, Esq., F.Z.S., Aug. 15, 1865.

663. *Chenalopex jubata* (Spix).   Orinoco Goose.
Hab. Tropical America.

   a. Purchased, April 27, 1865.

Subfamily CYGNINÆ.

Genus CYGNUS.

664. *Cygnus olor* (Gm.).   Common Swan.
Hab. Europe.

   a–c. Deposited, Feb. 26, 1866.

665. *Cygnus ferus,* Leach.   Hooper Swan.
Hab. Europe.

   a, b. Purchased, May 10, 1866.

666. *Cygnus buccinator,* Rich.   Trumpeter Swan.
Hab. North America.

   a–c. Received in exchange, April 21, 1866.

667. *Cygnus nigricollis* (Gm.).   Black-necked Swan.
Hab. Chili.

   a, b. Males.   Bred in the Gardens, June 23, 1858.
   c. Female.   Received in exchange, April 12, 1862.
   d–f. Bred in the Gardens, May 19, 1865.
   g. Bred in the Gardens, May 4, 1866.

668. *Cygnus atratus,* Lath.   Black Swan.
Hab. Australia.

   a. Male.   Purchased, Jan. 15, 1858.
   b. Female.   Presented by Dr. Mueller, C.M.Z.S., Sept. 19,
      1860.   From Melbourne.
   c. Bred in the Gardens, Nov. 11, 1864.
   d–g. Purchased, Oct. 19, 1865.
   h. Male.   Purchased, Oct. 27, 1865.
   i, j. Deposited, March 12, 1866.

Subfamily ANATINÆ.

Genus DENDROCYGNA.

669. *Dendrocygna viduata* (Linn.).   White-faced Tree-Duck.
Hab. Brazil.

   a–c. Presented by W. D. Christie, Esq., F.Z.S., May 28, 1862.
   d. Received in exchange, Oct. 15, 1863.

670. *Dendrocygna arcuata* (Cuv.).　　Indian Tree-Duck.
Hab. India.

a–c. Purchased, Sept. 3, 1858.
d. Presented by the Babu Rajendra Mullick, C.M.Z.S., July
14, 1857.
e, f. Purchased, Jan. 1, 1866.

671. *Dendrocygna arborea* (Linn.).　　Black-billed Tree-Duck.
Hab. Cuba.

a–d. Purchased, Dec. 26, 1863.

672. *Dendrocygna autumnalis* (Linn.).　　Red-billed Tree-
Duck.
Hab. Tropical America.

a–c. Purchased, March 8, 1864.　From Para.
d–g. Received in exchange, June 8, 1864.　From Mexico.

Genus TADORNA.

673. *Tadorna vulpanser*, Flem.　　Common Sheldrake.
Hab. Europe.

a. Male; b. Female.　Purchased, June 24, 1860.
c. Male; d. Female.　Hybrids between this species and *Casarca
cana* (Gm.).　Bred in the Gardens, June 26, 1859.

674. *Tadorna rutila* (Pall.).　　Ruddy Sheldrake.
Hab. Europe.

a. Male; b. Female.　Bred in the Gardens, May 13, 1859.
c. Male; d. Female.　Bred in the Gardens, June 2, 1861.
e. Male.　Received in exchange, Nov. 13, 1862.
f. Bred in the Gardens, 1863.

675. *Tadorna tadornoïdes*, Jard. & Selb.　　Australian Sheldrake.
Hab. South Australia.

a–c. Females.　Presented by the Hon. J. C. Hawker, Speaker
of the House of Assembly, Adelaide, April 3, 1862.
d, e. Males.　Presented by the Acclimatization Society of
Victoria, Sept. 22, 1863.
f–h. Presented by the Acclimatization Society of Victoria,
Jan. 4, 1865.

676. *Tadorna variegata* (Gm.). Variegated Sheldrake.
Hab. New Zealand.

> a. Male ; b. Female. Presented by J. D. Tetley, Esq., Aug.
> 3, 1863. Specimens figured, P. Z. S. 1864, pl. xviii.
> c–g. Bred in the Gardens, May 17, 1865.
> h–j. Bred in the Gardens, May 8, 1866.

Genus AIX.

677. *Aix sponsa* (Linn.). Summer Duck.
Hab. North America.

> a. Male ; b. Female. Bred in the Gardens, May 24, 1859.
> c. Male. Hybrid between this species and *Fuligula ferina*
> (Linn.).
> d, e. Males. Hybrids between this species and *Nyroca leu-
> cophthalma* (Bechst.). Received in exchange, Nov. 11,
> 1860.
> f–m. Bred in the Gardens, 1863.
> n–s. Deposited, 1863.
> t. Bred in the Gardens, May 23, 1865.
> u. Bred in the Gardens, June 9, 1865.
> v. Bred in the Gardens, June 18, 1866.

678. *Aix galericulata* (Linn.). Mandarin Duck.
Hab. China.

> a. Female. Bred in the Gardens, June 2, 1859.
> b. Male. Deposited, April 15, 1862.
> c. Male ; d. Female. Bred in the Gardens, 1862.
> e. Male. Bred in the Gardens, June 1, 1863.
> f. Male. Received in exchange, March 23, 1865.
> g–i. Bred in the Gardens, June 14, 1865.
> j–n. Bred in the Gardens, July 7, 1866.

Genus MARECA.

679. *Mareca penelope* (Linn.). Wigeon.
Hab. Europe.

> a. Male ; b. Female. Purchased, 1861.
> c. Male ; d. Female. Supposed hybrids between this species
> and the *Querquedula crecca* (Linn.), known as the Bima-
> culated Duck. Purchased, Feb. 18, 1861.
> e, f. Supposed hydrids between this species and the *Querque-
> dula crecca* (Linn.), known as the Bimaculated Duck.
> Presented by J. A. Heaton, Esq., Oct. 18, 1864.

174 ANATIDÆ.

Genus DAFILA.

680. *Dafila acuta* (Linn.). Pintail.
Hab. Europe.

a, b. Males; *c.* Female. Bred in the Gardens, May 15, 1861.
d–f. Females. Bred in the Gardens, May 16, 1860.
g–i. Males. Purchased, May 22, 1865.

Genus PŒCILONETTA.

681. *Pœcilonetta bahamensis* (Linn.). Bahama Duck.
Hab. West Indies.

a. Received in exchange, Dec. 3, 1860.
b–d. Bred in the Gardens, July 31, 1860.
e. Deposited, Aug. 16, 1862.
f. Male. Deposited, 1863.
g–i. Bred in the Gardens, June 1, 1863.
j–m. Bred in the Gardens, June 14, 1865.
n–t. Bred in the Gardens, July 6, 1865.
u. Bred in the Gardens, July 7, 1866.

682. *Pœcilonetta erythrorhyncha* (Gm.). Red-billed Duck.
Hab. South Africa.

a. Male; *b.* Female. Bred in the Gardens, July 9, 1860.

Genus ANAS.

683. *Anas obscura*, Gm. Dusky Duck.
Hab. North America.

a. Male; *b.* Female. Bred in the Gardens, May 21, 1861.
c. Bred in the Gardens, 1860.
d, e. Presented by A. Downs, Esq., C.M.Z.S., Oct. 4, 1863.
From Halifax.
f. Bred in the Gardens, May 23, 1865.
g–j. Bred in the Gardens, May 8, 1866.
k–l. Presented by A. Downs, Esq., C.M.Z.S., Oct. 7, 1866.
From Halifax, Nova Scotia.

684. *Anas xanthorhyncha*, Forst. Yellow-billed Duck.
Hab. South Africa.

a. Male; *b.* Female. Bred in the Gardens, May 20, 1859.
c. Male; *d.* Female. Bred in the Gardens, May 30, 1860.
e, f. Bred in the Gardens, June 1, 1863.

685. *Anas superciliosa*, Gm.   Australian Wild Duck.
Hab. Australia.

   a. Female.   Presented by Edward Wilson, Esq., May 31,
       1860.   From Melbourne.
   b, c. Presented by Dr. Mueller, C.M.Z.S., March 18, 1863.
       From Melbourne.
   d–i. Presented by Dr. Mueller, C.M.Z.S., Sept. 5, 1865.
   j–t. Deposited, July 20, 1866.
   u–z. Presented by the Acclimatization Society of Melbourne,
       July 21, 1866.

686. *Anas punctata*, Cuv.   Chestnut-breasted Duck.
Hab. Australia.

   a. Female.   Purchased, May 11, 1865.
   b, c. Deposited, July 20, 1866.

687. *Anas strepera*, Linn.   Common Gadwall.
Hab. Europe.

   a, b. Males ; *c. d.* Females.   Bred in the Gardens, June 20,
       1861.

Genus QUERQUEDULA.

688. *Querquedula crecca* (Linn.).   Common Teal.
Hab. Europe.

   a. Male ; *b.* Female.   Bred in the Gardens, June 24, 1860.
   c. Male ; *d.* Female.   Bred in the Gardens, June 1861.
   e, f. Females.   Purchased, May 22, 1865.
   g–i. Bred in the Gardens, July 6, 1865.
   j. Presented by T. M. Hayward, Esq., June 28, 1866.

689. *Querquedula brasiliensis* (Gm.).   Brazilian Teal.
Hab. Para.

   a. Purchased, March 8, 1864.

690. *Querquedula circia* (Linn.).   Garganey Teal.
Hab. Europe.

   a. Male.   Presented by Earl Fitzwilliam, F.Z.S., Feb. 11,
       1858.
   b, c. Males ; *d, e.* Females.   Purchased, May 22, 1865.

Genus SPATULA.

691. *Spatula clypeata* (Linn.). Shoveller.
Hab. Europe.

a–c. Males. Bred in the Gardens, July 4, 1859.
d, e. Males ; *f, g.* Females. Purchased, May 22, 1865.

Subfamily FULIGULINÆ.

Genus FULIGULA.

692. *Fuligula cristata* (Ray). Tufted Duck.
Hab. Europe.

a, b. Males ; Purchased, Jan. 16, 1862.
c, d. Females. Purchased, Nov. 22, 1860.
e–g. Hybrids between this species and *Nyroca leucophthalma*
(Bechst.), for three or four generations. Bred in the
Gardens, June 12, 1861.
h, i. Males ; *j, k.* Females, Purchased, May 22, 1865.

693. *Fuligula marila* (Linn.). Scaup Duck.
Hab. Europe.

a. Male ; *b.* Female. Purchased, April 24, 1861.
c. Hybrid, supposed to be between this species and *Nyroca
leucophthalma* (Bechst.). Purchased, March 26, 1861.
d–i. Males ; *j–o.* Females. Received in exchange, March 10,
1865.

694. *Fuligula ferina* (Linn.). Red-headed Pochard.
Hab. Europe.

a. Male : *b.* Female. Purchased, Nov. 29, 1860.

Genus NYROCA.

695. *Nyroca leucophthalma* (Bechst.). White-eyed or Cas-
taneous Duck.
Hab. Europe.

a. Male. Purchased, 1857.
b, c. Males ; *d–g.* Females. Purchased, May 22, 1865.

Genus CLANGULA.

696. *Clangula glaucion* (Linn.). Golden-eye.
Hab. Europe.

a–i. Purchased, Jan. 16, 1862.
j. Female. Purchased, March 3, 1862.
k–s. Received in exchange, April 6, 1866.

Genus ŒDEMIA.

697. *Œdemia nigra*, Flem. Common or Black Scoter.
Hab. British Islands.
a. Purchased, May 11, 1864.
b–e. Purchased, May 10, 1866.

Genus SOMATERIA.

698. *Somateria mollissima* (Linn.). Eider Duck.
Hab. Europe.

a, b. Purchased, Nov. 2, 1866.

Genus MERGUS.

699. *Mergus merganser*, Linn. Goosander.
Hab. British Islands.

a, b. Females. Presented by Anthony Sanaze, Esq., March 5,
1864.
c. Male. Purchased, Nov. 28, 1865.

700. *Mergus serrator*, Linn. Red-breasted Merganser.
Hab. British Islands.

a, b. Purchased, Aug. 13, 1866.

Family PELECANIDÆ.

Genus PELECANUS.

701. *Pelecanus onocrotalus*, Linn. White Pelican.
Hab. Southern Europe and North Africa.

a. Purchased, April 22, 1851. From Egypt.
b. Purchased, July 9, 1852. From Egypt.
c. Purchased, April 12, 1853. From Egypt.

N

702. *Pelecanus rufescens*, Lath.   Red-backed Pelican.
    Hab. West Africa.

    a. Purchased, May 18, 1866.

703. *Pelecanus crispus*, Feldegg.   Crested Pelican.
    Hab. Southern Europe.

    a. Purchased, Sept. 26, 1853.

704. *Pelecanus fuscus*, Linn.   Brown Pelican.
    Hab. West Indies.

    a, *b*. Presented by Capt. Abbott, July 18, 1854.

705. *Pelecanus conspicillatus*, Gould.   Australian Pelican.
    Hab. Australia.

    a. Purchased, April 10, 1864.
    b, *c*. Presented by the Acclimatization Society of Victoria,
        June 27, 1864.

Genus SULA.

706. *Sula bassana*, Linn.   Gannet.
    Hab. British Islands.

    a. Presented by J. J. Broadwood, Esq., July 13, 1864.
    b. Deposited, April 13, 1866.

Genus PHALACROCORAX.

707. *Phalacrocorax carbo*, Linn.   Common Cormorant.
    Hab. British Islands.

    a. Purchased, Sept. 26, 1853.   From Egypt.
    b. Presented by Sir Henry Stracey, Bart., F.Z.S., Dec. 1, 1852.

708. *Phalacrocorax lugubris*, Rüpp.   West African Cormorant.
    Hab. West Africa.

    a. Purchased, Aug. 19, 1865.

Family LARIDÆ.
Genus LARUS.

709. *Larus marinus*, Linn.   Greater Black-backed Gull.
    Hab. British Islands.

    a. Male.   Presented by W. H. Leach, Esq., June 3, 1861.

b. Female. Presented by W. N. Turner, Esq., Oct. 27, 1848.
c. Deposited, 1861.

710. *Larus glaucus,* Linn. Glaucous Gull.
Hab. Europe.

a. Purchased, Dec. 23, 1859.
b. Purchased, Oct. 3, 1860.

711. *Larus fuscus,* Linn. Lesser Black-backed Gull.
Hab. British Islands.

a. Presented by Mrs. Cotton, 1861.

712. *Larus fuscescens,* Licht. Mediterranean Gull.
Hab. North Africa.

a. Purchased, Aug. 3, 1859. From Mogador*.

713. *Larus argentatus,* Brünn. Herring Gull.
Hab. Europe.

a. Presented by S. Redman, Esq., Feb. 4, 1860.
b. Presented by — Tomkinson, Esq., Oct. 25, 1861.
c. Presented by Thomas Page, Esq., Nov. 12, 1862.
d. Bred in the Gardens, 1864.
e, f. Presented by H. Houlder, Esq., Sept. 12, 1864.
g, h. Presented by Thomas Walker, Esq., F.Z.S., Feb. 11, 1865.
i. Presented by R. Tate, Esq., Aug. 5, 1865.

714. *Larus ridibundus,* Linn. Black-headed Gull.
Hab. Europe.

a. Presented by J. Salmon, Esq., July 25, 1856.
b. Presented by Dr. Bree, May 1861.
c. Presented by Alfred Newton, Esq., F.Z.S., July 21, 1851.
d. Deposited, April 30, 1866. From Norfolk.

Genus RISSA.

715. *Rissa tridactyla* (Linn.). Kittiwake Gull.
Hab. British Islands.

a. Purchased, April 28, 1865.

Genus STERNA.

716. *Sterna hirundo,* Linn. Common Tern.
Hab. British Islands.

a. Purchased, Sept. 13, 1866.

---

* See Mr. Sclater's remarks on this bird, P. Z. S., 1867, p. 315.

Genus Anous.

717. *Anous stolidus* (Linn.).   Noddy Tern.
Hab. West Indies.

    a. Purchased, May 18, 1865.

Family COLYMBIDÆ.

Genus Podiceps.

718. *Podiceps minor*, Lath.   Little Grebe.
Hab. British Islands.

    a. Purchased, Feb. 21, 1863.
    b–e. Presented by Master A. M. Hall, Dec. 1, 1863.

719. *Podiceps rubricollis*, Lath.   Red-necked Grebe.
Hab. British Islands.

    a. Presented by W. C. Horsfall, Esq., Feb. 16, 1866.

Family ALCIDÆ.

Genus Alca.

720. *Alca torda*, Linn.   Razor-bill.
Hab. British Islands.

    a. Presented by P. W. Symonds, Esq., July 28, 1864.

Family SPHENISCIDÆ.

Genus Apterodytes.

721. *Apterodytes pennantii,* Gray.   Pennant's Penguin.
Hab. Falkland Islands.

    a. Presented by Commander Fenwick, R.N., March 27, 1865.

Class REPTILIA.

Order TESTUDINATA.

Family TESTUDINIDÆ.

Genus TESTUDO.

1. *Testudo sulcata*, Miller.   Grooved Tortoise.
   Hab. Galam, Africa.

   a. Purchased, July 6, 1862.

2. *Testudo pardalis*, Bell.   Leopard Tortoise.
   Hab. Africa.

   a, b. Presented by Sir William Williams, Bart., F.Z.S., Oct. 19,
   1866.

3. *Testudo tabulata*, Walb.   Brazilian Tortoise.
   Hab. British Guiana.

   a. Presented by D. C. Munro, Esq., H.B.M.'s Consul, Surinam,
   May 19, 1863.
   b, c. Purchased, May 30, 1865.

4. *Testudo radiata*, Shaw.   Radiated Tortoise.
   Hab. Madagascar.

   a–e. Deposited, Aug. 11, 1863.

5. *Testudo indica*, Gm.   Indian Tortoise.
   Hab. Seychelles.

   a, b. Presented by the Babu Rajendra Mullick, C.M.Z.S., July
   25, 1864.

Family EMYDIDÆ.

Genus EMYS.

6. *Emys guttata*, Schweig.   Speckled Terrapen.
   Hab. North America.

   a–c. Presented by A. Downs, Esq., C.M.Z.S., Aug. 1, 1862.
   From Halifax.

7. *Emys rubriventris*, Lecontc.   Red-bellied Terrapen.
   Hab. Potomac River, North America.

   a. Presented by the Smithsonian Institution, Washington,
   U.S.A., May 22, 1866.

8. *Emys picta*, Schw.   Painted Terrapen.
   Hab. Potomac River, North America.

   a–i. Presented by the Smithsonian Institution, Washington,
   U.S.A., May 29, 1866.
   j. Presented by A. Downs, Esq., C.M.Z.S., Oct. 7, 1866.   From
   Halifax, U.S.A.

9. *Emys terrapin*, Schœpff.   Salt-water Terrapen.
   Hab. Potomac River, North America.

   a, b. Presented by the Smithsonian Institution, Washington,
   U.S.A., May 29, 1866.

Genus CHELYDRA.

10. *Chelydra serpentina* (Linn.).   Alligator Terrapen.
    Hab. North America.

    a, b. Presented by Arthur Russell, Esq., F.Z.S., June 24, 1860.
    c. Presented by A. Downs, C.M.Z.S., Oct. 7, 1866.   From
    Halifax, U.S.A.

Genus DERMATEMYS.

11. *Dermatemys mawii*, Gray.   Maw's Terrapen.
    Hab. South America.

    a. Purchased.

Family CHELYDIDÆ.

Genus CHELODINA.

12. *Chelodina oblonga*, Gray.   Oblong Chelodine.
    Hab. West Australia.

    a. Presented by W. Ayshford Sandford, Esq., March 18, 1856.

13. *Chelodina longicollis* (Shaw).   Long-necked Chelodine.
    Hab. River Yarra, Australia.

    a. Presented by P. Joske, Esq., Jan. 9, 1861.

Family CHELONIIDÆ.

Genus CARETTA.

14. *Caretta imbricata* (Schweig.).   Hawk's-billed Turtle.
    Hab. East Indies.

    a. Presented by Messrs. T. C. W. Mackay & Co., July 27, 1864.

Genus CHELONIA.

15. *Chelonia viridis* (Schneid.).   Common Turtle.
    Hab. West Indies.

    a. Presented by C. Butler, Esq., Sept. 2, 1864.
    b, c. Young.   Presented by H. Jones, Esq.   From Ascension
       Island.

Order CROCODILIA.

Family CROCODILIDÆ.

Genus ALLIGATOR.

16. *Alligator mississippiensis* (Daud.).   Alligator.
    Hab. Mississippi.

    a, b. Purchased, 1863.
    c–e. Purchased, Oct. 9, 1864.
    f. Presented by J. Oppenheim, Esq., May 9, 1865.

Genus CROCODILUS.

17. *Crocodilus*, sp.   Crocodile.
    Hab. Africa (?).

    a. Purchased, July 13, 1866.

Order SAURIA.

Family MONITORIDÆ.

Genus MONITOR.

18. *Monitor niloticus*, Hasselq.   Egyptian Monitor.
    Hab. North Africa.

    a. Purchased, Sept. 28, 1863.

19. *Monitor albogularis* (Gray).   White-throated Monitor.
    Hab. Africa.

    a. Purchased, Feb. 23, 1865.

20. *Monitor gouldi*, Schleg.   Australian Monitor.
    Hab. Australia.

    a. Purchased, March 9, 1865.
    b, c. Purchased, May 11, 1865.
    d. Presented by Dr. Mueller, C.M.Z.S., May 26, 1865.
    e. Presented by the Acclimatization Society of Melbourne,
       July 21, 1866.

Genus HYDROSAURUS.

21. *Hydrosaurus bivittatus*, Wagl.   Two-banded Lizard.
    Hab. Africa (?).

    a. Presented by P. F. Debary, Esq., F.Z.S., Oct. 25, 1866.

Family TEIIDÆ.

Genus TEIUS.

22. *Teius teguexin* (Linn.).   Teguexin Lizard.
    Hab. South America.

    a. Purchased, June 5, 1865.

Genus AMEIVA.

23. *Ameiva dorsalis*, Gray.   Dorsal Lizard.
    Hab. St. Thomas.

    a. Presented by Capt. Abbott, April 28, 1864.

Family LACERTIDÆ.

Genus LACERTA.

24. *Lacerta viridis*, Linn.   Green Lizard.
    Hab. Island of Jersey.

    a. Purchased, April 13, 1864.
    b. Presented by — Baker, Esq., Sept. 5, 1864.
    c–h. Purchased, May 19, 1865.

25. *Lacerta agilis,* Linn.   Sand-Lizard.
Hab. Europe.

> *a.* Presented by F. Coleman, Esq., April 30, 1865.
> *b–k.* Presented by the Rev. C. Wolley, April 9, 1866.

26. *Lacerta ocellata,* Daud.   Eyed Lizard.
Hab. South Europe.

> *a–j.* Presented by J. Braxton Hicks, M.D., Oct. 22, 1866.

27. *Lacerta vivipara.*   Common Lizard.
Hab. England.

> *a–d.* Purchased, 1866.

Family ZONURIDÆ.

Genus PSEUDOPUS.

28. *Pseudopus pallasii* (Oppel).   Glass Snake.
Hab. Dalmatia.

> *a–d.* Purchased, 1854.

Family SCINCIDÆ.

Genus TRACHYDOSAURUS.

29. *Trachydosaurus rugosus,* Gray.   Stump-tailed Lizard.
Hab. New Holland.

> *a.* Purchased, 1858.
> *b.* Presented by the Rev. W. H. Hawker, F.Z.S., April 4, 1862.

Genus TROPIDOLEPISMA.

30. *Tropidolepisma majus,* Gray.   Large Tropidolepisma.
Hab. New South Wales.

> *a, b.* Presented by George Macleay, Esq., F.Z.S., May 19, 1862.

Genus CYCLODUS.

31. *Cyclodus gigas* (Bodd.).   Great Cyclodus.
Hab. Australia.

> *a.* Purchased, 1856.

b. Presented by the Acclimatization Society of Victoria, March
    31, 1864.
c. Purchased, April 19, 1866.
d–m. Bred in the Gardens, July 10, 1866.

Genus EGERNIA.

32. *Egernia cunninghamii,* Gray.   Cunningham's Skink.
    Hab. Australia.

    a. Presented by the Acclimatization Society of Victoria, March
        31, 1864.

Genus GONGYLUS.

33. *Gongylus ocellatus,* Forsk.   Ocellated Skink.
    Hab. South Europe.

    a. Purchased, Aug. 8, 1863.

Genus EUPREPES.

34. *Euprepes australis,* Gray.   Australian Skink.
    Hab. Australia.

    a–o. Received, July 19, 1866.

Family IGUANIDÆ.

Genus IGUANA.

35. *Iguana tuberculata,* Laur.   Tuberculated Lizard.
    Hab. St. Thomas.

    a. Purchased, Oct. 29, 1864.
    b. Presented by Capt. Sawyer, Nov. 15, 1864.
    c. Purchased, June 30, 1865.
    d. Presented by Edward Greey, Esq., F.Z.S., Oct. 29, 1866.

Genus ANOLIS.

36. *Anolis cristatellus,* Dum. et Bibr.   Crested Anolis.
    Hab. West Indies.

    a–j. Presented by Capt. Sawyer, Sept. 14, 1864.
    k, l. Deposited, Sept. 30, 1864.

Family AGAMIDÆ.

Genus Amphibolurus.

37. *Amphibolurus barbatus* (Cuv.). Bearded Lizard.
Hab. Australia.

    a. Purchased, March 14, 1864.
    b. Purchased, June 15, 1866.

Genus Moloch.

38. *Moloch horridus*, Gray. Moloch Lizard.
Hab. Australia.

    a. Presented by S. S. Travers, Esq., June 11, 1866.
    b. Purchased, Aug. 20, 1866.

Family CHAMÆLEONIDÆ.

Genus Chamæleon.

39. *Chamæleon vulgaris*, Daud. Common Chameleon.
Hab. North Africa.

    a. Deposited, Dec. 10, 1863.
    b. Presented by W. Gass, Esq., Aug. 26, 1863.
    c. Presented by B. E. Spraull, Esq., Dec. 28, 1863. From China.
    d. Presented by T. Stillwell, Esq., Jan. 5, 1865.
    e. Presented by J. Bramley, Esq., July 30, 1865.

Family GECCOTIDÆ.

Genus Gecco.

40. *Gecco verus*, Merr. Indian Gecko.
Hab. India.

    a–e. Presented by Capt. Frain, July 24, 1866.

Order OPHIDIA.

Family BOIDÆ.

Genus PYTHON.

41. *Python sebæ* (Gm.).   West African Python.
Hab. West Africa.

> *a*. Male.   Purchased, April 18, 1859.
> *b*. Presented by Dr. Marchisio, Dec. 18, 1863.
> *c*. Purchased, July 21, 1864.
> *d*. Purchased, Jan. 17, 1865.
> *e, f*. Purchased, July 4, 1865.
> *g*. Presented by Dr. W. A. Gardiner, Aug. 18, 1865.

42. *Python regius* (Shaw).   Royal Python.
Hab. West Africa.

> *a*. Purchased, April 8, 1859.

43. *Python molurus* (Linn.).   Indian Python.
Hab. India.

> *a*. Purchased, Sept. 24, 1863.
> *b*. Presented by Capt. A. Henley, 52nd Regt., Aug. 17, 1866.
> *c*. Received, Aug. 17, 1866.

Genus MORELIA.

44. *Morelia spilotes* (Lacép.).   Diamond Snake.
Hab. New South Wales.

> *a, b*. Presented by Gerard Krefft, Esq., C.M.Z.S., Sept. 29, 1863.
> *c*. Presented by the Acclimatization Society of Victoria, March
> 31, 1864.
> *d*. Presented by Gerard Krefft, Esq., C.M.Z.S., July 21, 1866.

Genus BOA.

45. *Boa constrictor*, Linn.   Common Boa.
Hab. South America.

> *a*. Presented by Hippesley Justins, Esq., July 5, 1861.
> *b*. Presented by J. R. Perry, Esq., H.B.M.'s Consul at Para,
> June 9, 1863.   From the Lower Amazon.
> *c*. Presented by Dr. Leard, April 30, 1864.

d. Presented by R. H. Soulter, Esq., June 15, 1865.
e. Presented by S. Lambert, Esq., Oct. 30, 1865.
f. Presented by Comm. R. M. Blomfield, R.N., H.M.S. 'Tamar,' July 3, 1866.
g, h. Purchased, Oct. 15, 1866. From Mexico.
i. Presented by Lieut. C. Balfour, R.N., H.M.S. 'Buzzard,' Nov. 17, 1866.

Genus EUNECTES.

46. *Eunectes murinus* (Linn.). Anaconda.
Hab. South America.

a. Purchased, Feb. 8, 1864.
b, c. Purchased, Oct. 4, 1865.

Genus EPICRATES.

47. *Epicrates angulifer*, Bibr. Pale-headed Snake.
Hab. Cuba.

a–d. Purchased, Jan. 25, 1865.

Genus CHILOBOTHRUS.

48. *Chilobothrus inornatus*, Dum. Yellow Snake.
Hab. Jamaica.

a, b. Presented by Dr. Bowerbank, Oct. 17, 1855.
c–e. Presented by Capt. Abbott, June 14, 1864.
f. Presented by Capt. Hammack, April 13, 1865.

Family ERYCIDÆ.

Genus ERYX.

49. *Eryx jaculus*, Daud. Eryx Snake.
Hab. Egypt.

a. Purchased, Aug. 8, 1863.

50. *Eryx johnii* (Russell). Clothonia.
Hab. India.

a. Purchased, Jan. 10, 1865.

Family TROPIDONOTIDÆ.

Genus TROPIDONOTUS.

51. *Tropidonotus natrix* (Linn.).   Common Snake.
    Hab. British Islands.

    a–f. Purchased.

52. *Tropidonotus quincunciatus* (Schleg.).   Common River-
        Snake.
    Hab. India.

    a. Purchased, July 4, 1861.

53. *Tropidonotus ordinatus*, Linn.   Garter Snake.
    Hab. Nova Scotia.

    a–d. Presented by A. Downs, Esq., C.M.Z.S., Oct. 2, 1863.

54. *Tropidonotus viperinus* (Merr.).   Viperine Snake.
    Hab. North Africa.

    a. Purchased.

Genus HETERODON.

55. *Heterodon madagascariensis,* Dum. & Bibr.   Sharp-snouted
        Snake.
    Hab. Madagascar.

    a. Presented by Edward Newton, Esq., C.M.Z.S., Nov. 27, 1863.

Family HERPETODRYADIDÆ.

Genus DROMICUS.

56. *Dromicus antillensis,* Schleg.   Antillean Snake.
    Hab. St. Thomas's, West Indies.

    a, b. Presented by Capt. Abbott, April 28, 1864.
    c. Presented by Edward Greey, Esq., F.Z.S., March 17, 1865.

Genus CYCLOPHIS.

57. *Cyclophis vernalis,* DeKay.   Grass Snake.
    Hab. Nova Scotia.

    a. Presented by A. Downs, Esq., C.M.Z.S., Oct. 2, 1863.

Family COLUBRIDÆ.

Genus CORONELLA.

58. *Coronella lævis*, Lacép.　Smooth Snake.
Hab. England.

 a. Purchased, July 2, 1863.
 b. Presented by John Pares, Esq., Sept. 13, 1864.
 c. Purchased, Sept. 7, 1865.　From Germany.
 d. Presented by the Rev. C. Wolley, April 9, 1866.

Genus COLUBER.

59. *Coluber quadrilineatus* (Pall.).　Four-lined Snake.
Hab. Egypt.

 a, b. Purchased, Aug. 8, 1863.

Genus PTYAS.

60. *Ptyas mucosa* (Linn.).　Rat-Snake.
Hab. India.

 a. Presented by the Babu Rajendra Mullick, C.M.Z.S., July 14, 1857.
 b. Presented by Dr. Shortt, F.Z.S., Oct. 23, 1863.

Genus ZAMENIS.

61. *Zamenis atrovirens*, Shaw.　Dark-green Snake.
Hab. Dalmatia.

 a. Purchased, Dec. 29, 1862.
 b. Purchased, Aug. 8, 1863.

62. *Zamenis cliffordii* (Schleg.).　Clifford's Snake.
Hab. Egypt.

 a. Purchased, Aug. 8, 1863.

63. *Zamenis hippocrepis*, Linn.　Horseshoe-Snake.
Hab. Germany.

 a. Purchased, Sept. 7, 1865.

Family ELAPIDÆ.

Genus NAIA.

64. *Naia tripudians* (Merr.).   Indian Cobra.
Hab. India.

    a. Purchased, May 24, 1862.

Family VIPERIDÆ.

Genus CENCHRIS.

65. *Cenchris piscivorus*, Gray.   Water-Viper.
Hab. North America.

    a–e. Received in exchange, 1858.
    f. Bred in the Gardens, Jan. 1865.

Genus VIPERA.

66. *Vipera ammodytes*, Klein.   Sand-Asp.
Hab. Egypt.

    a, b. Purchased, Aug. 8, 1863.
    c. Purchased, Sept. 7, 1865.   From Germany.

67. *Vipera aspis* (Linn.).   Black Viper.
Hab. Germany.

Genus PELIAS.

68. *Pelias berus*, Merr.   Common Adder.
Hab. British Islands.

    a. Presented by F. T. Buckland, Esq., F.Z.S., May 11, 1864.
    b. Presented by E. J. Lowe, Esq , June 28, 1864.
    c, d. Presented by Thomas Barling, Esq., April 29, 1865.

Genus CLOTHO.

69. *Clotho arietans*, Gray.   Puff-Adder.
Hab. Cape of Good Hope.

    a. Purchased, Aug. 14. 1862.

70. *Clotho rhinoceros,* Schl.　River-Jack.
　　Hab. West Africa.
　　　a. Presented by George Atkins, Esq., Oct 23, 1865.

71. *Clotho cornuta* (Daud.).　Horned Viper.
　　Hab. South Africa.

　　　a–d. Purchased, July 20, 1865.

Family CROTALIDÆ.

Genus CROTALUS.

72. *Crotalus durissus* (Daud.).　Rattle-Snake.
　　Hab. North America.

　　　a. Male; *b.* Female.　Purchased, July 3, 1862.

Genus CRASPEDOCEPHALUS.

73. *Craspedocephalus bilineatus,* Wied.　Two-lined Palm-
　　Snake.
　　Hab. Brazil.

　　　a. Presented by Dr. Wucherer, C.M.Z.S., Oct. 1, 1864.　From
　　Bahia.

Class BATRACHIA.

Family BUFONIDÆ.

Genus Bufo.

1. *Bufo pantherinus*, Geoff. St.-Hil.   Pantherine Toad.
   Hab. Tunis.

   a. Presented by P. L. Sclater, Esq., F.Z.S., Secretary of the Society, March 24, 1859.

2. *Bufo calamita*, Laur.   Natterjack Toad.
   Hab. Cornwall.

   a. Presented by Dr. Lankester, Aug. 1861.
   b, c. Presented by — Tait, Esq., April 19, 1865.

3. *Bufo vulgaris*, Laur.   Common Toad.
   Hab. England.

   a–e. Presented, 1860.

Family RANIDÆ.

Genus Rana.

4. *Rana esculenta*, Linn.   Edible Frog.
   Hab. Europe.

   a, b. Purchased, 1862.

5. *Rana halecina*, Catesby.   American Green Frog.
   Hab. Nova Scotia.

   a, b. Presented by A. Downs, Esq., C.M.Z.S., July 9, 1861. From Halifax.

6. *Rana clamata*, Daud.   Noisy Frog.
   Hab. Nova Scotia.

   a. Presented by A. Downs, Esq., C.M.Z.S., July 9, 1861.   From Halifax, U.S.A.

7. *Rana mugiens*, Merr.   Bull Frog.
   Hab. Nova Scotia.

   a. Presented by A. Downs, Esq., C.M.Z.S., July 9, 1861.   From Halifax.

Genus CYSTIGNATHUS.

8. *Cystignathus ocellatus.* West-Indian Frog.
Hab. West Indies.

   a–h. Purchased, Aug. 20, 1866.

Family BOMBINATORIDÆ.

Genus BOMBINATOR.

9. *Bombinator igneus.* Fire-bellied Toad.
Hab. Europe.

   a–j. Purchased, Aug. 24, 1865.
   k–p. Purchased, Sept. 24, 1866.

Genus PSEUDOPHRYNE.

10. *Pseudophryne australis* (Gray). Australian Frog.
Hab. New South Wales.

   a. Presented by Gerard Krefft, Esq., C.M.Z.S., Sept 29, 1863.

Family HYLIDÆ.

Genus HYLA.

11. *Hyla arborea* (Linn.). European Tree-Frog.
Hab. Europe.

   a, b. Purchased, May 12, 1862.

12. *Hyla ewingii,* Dum. et Bibr. Ewing's Tree-Frog.
Hab. New South Wales.

   a. Purchased, April 23, 1863.

13. *Hyla citropus,* Dum. et Bibr. Tree-Frog.
Hab. New South Wales.

   a. Purchased, April 23, 1863.

14. *Hyla peronii,* Dum. et Bibr. Péron's Tree-Frog.
Hab. New South Wales.

   a, b. Purchased, April 23, 1863.

15. *Hyla kreffti*, Günth.   Krefft's Tree-Frog.
    Hab. New South Wales.

    a, b. Purchased, April 23, 1863.

16. *Hyla phyllochroa*, Günth.   Leaf-green Tree-Frog.
    Hab. New South Wales.

    a. Purchased, April 23, 1863.

Genus PELODRYAS.

17. *Pelodryas cærulea* (White).   White's Tree-Frog.
    Hab. New South Wales.

    a. Presented by George Macleay, F.Z.S., May 19, 1862.

Family SALAMANDRIDÆ.

Genus SALAMANDRA.

18. *Salamandra maculosa* (Linn.).   Spotted Salamander.
    Hab. Europe.

    a. Presented by Mrs. Hopper, June 27, 1861.
    b–f. Purchased, June 2, 1863.
    g, h. Purchased, Aug. 24, 1865.
    i–n. Purchased, Sept. 24, 1866.

Genus TRITON.

19. *Triton cristatus*, Linn.   Crested Newt.
    Hab. British Islands.

    a–c. Presented by T. C. Eyton, Esq., F.Z.S., April 14, 1862.

20. *Triton punctatus*, Dugès.   Smooth Newt.
    Hab. British Islands.

    a–c. Presented by T. C. Eyton, Esq., F.Z.S., April 14, 1862.

21. *Triton palmatus*, Otth.   Palmated Newt.
    Hab. British Islands.

    a–c, &c. Presented by E. W. H. Holdsworth, Esq., F.Z.S., 1863.

Genus AMBLYSTOMA.

22. *Amblystoma luridum*, Baird.   Illinois Salamander.
    Hab. Illinois, U.S.A.

    a, b. Presented by the Smithsonian Institution, Dec. 1857.

Genus SIREDON.

23. *Siredon mexicanus* (Shaw).   The Axolotl.
    Hab. Mexico.

    a, b. Purchased, Aug. 30, 1864.
    c–n. Purchased, May 8, 1866.

Family PROTONOPSIDÆ.

Genus SIEBOLDIA.

24. *Sieboldia maxima* (Schleg.).   Gigantic Salamander.
    Hab. Japan.

    a. Purchased, March 12, 1860.
    b. Deposited by Capt. Taylor, July 2, 1861.
    c, d. Purchased, June 19, 1862.
    e. Deposited, Feb. 27, 1865.

Family PROTEIDÆ.

Genus PROTEUS.

25. *Proteus anguinus* (Shaw).   Proteus.
    Hab. Europe.

    a. Presented by Dr. Percival Wright.
    b, c. Presented by T. H. Chambers, Esq., July 17, 1863.
    d. Presented by F. M. Burton, Esq., Nov. 17, 1863.

Class PISCES*.

Family LEPIDOSIRENIDÆ.

Genus PROTOPTERUS.

1. *Protopterus annectens*, Owen.   African Lepidosiren.
   Hab. River Gambia.

   a, b. Purchased, May 24, 1865.
   c. Purchased, May 27, 1865.
   d. Purchased, Oct. 4, 1865.

Family SQUALIDÆ.

Genus SCYLLIUM.

2. *Scyllium canicula* (Linn.).   Small-spotted Dogfish.
   Hab. British Seas.

   a. Purchased, Sept. 25, 1863.

Family ACIPENSERIDÆ.

Genus ACIPENSER.

3. *Acipenser sturio*, Linn.   Common Sturgeon.
   Hab. European rivers.

   a. Purchased, July 26, 1863.
   b. Small example from the Elbe, purchased, 1864.

Family GASTEROSTEIDÆ.

Genus GASTEROSTEUS.

4. *Gasterosteus spinachia*, Linn.   Fifteen-spined Stickleback.
   Hab. British Seas.

   * In this class it is not generally possible to keep a record of
individual specimens; but the names of all the determinable species exhibited in the Gardens since Jan. 1st, 1862, are given
in their natural order.—P. L. S.

5. *Gasterosteus trachurus*, Cuv. et Val. Rough-tailed Stickleback.
   Hab. British Seas.

6. *Gasterosteus leiurus*, Cuv. et Val. Smooth-tailed Stickleback.
   Hab. British Seas.

7. *Gasterosteus pungitius*, Linn. Ten-spined Stickleback.
   Hab. British Seas.

Family PERCIDÆ.

Genus PERCA.

8. *Perca fluviatilis*, Linn. Common Perch.
   Hab. British fresh waters.

Genus ACERINA.

9. *Acerina cernua* (Linn.). Ruffe, or Pope.
   Hab. British waters.

Genus LABRAX.

10. *Labrax lupus* (Linn.). Bass.
    Hab. British Seas.

Family TRIGLIDÆ.

Genus COTTUS.

11. *Cottus quadricornis*, Linn. Four-horned Cottus.
    Hab. British Seas.

12. *Cottus bubalis*, Euphr. Long-spined Cottus.
    Hab. British Seas.

Genus ASPIDOPHORUS.

13. *Aspidophorus cataphractus* (Linn.). Armed Bullhead.
    Hab. British Seas.

Family MUGILIDÆ.

Genus MUGIL.

14. *Mugil capito,* Cuv.   Grey Mullet.
    Hab. British Seas.

Family BLENNIIDÆ.

Genus CENTRONOTUS.

15. *Centronotus gunnellus* (Linn.).   Spotted Gunnel.
    Hab. British Seas.

Genus ZOARCES.

16. *Zoarces viviparus,* Cuv.   Viviparous Blenny.
    Hab. British Seas.

Genus BLENNIUS.

17. *Blennius pholis,* Linn.   Smooth Shanny.
    Hab. British Seas.

Family GOBIIDÆ.

Genus GOBIUS.

18. *Gobius niger,* Linn.   Black Goby.
    Hab. British Seas.

19. *Gobius bipunctatus,* Yarr.   Two-spotted Goby.
    Hab. British Seas.

20. *Gobius minutus,* Pall.   Freckled Goby.
    Hab. British Seas.

Genus CALLIONYMUS.

21. *Callionymus dracunculus* (Linn.).   Sordid Dragonet.
    Hab. British Seas.

Family LABRIDÆ.

Genus CRENILABRUS.

22. *Crenilabrus cornubicus*, Don.   Goldfinny.
Hab. British Seas.

Genus CTENOLABRUS.

23. *Ctenolabrus rupestris* (Linn.).   Jago's Goldfinny.
Hab. British Seas.

Genus LABRUS.

24. *Labrus maculatus*, Bloch.   Green-streaked Wrasse.
Hab. British Seas.

Family PLEURONECTIDÆ.

Genus PLATESSA.

25. *Platessa flesus* (Linn.).   Flounder.
Hab. British Seas.

26. *Platessa vulgaris*, Flem.   Plaice.
Hab. British Seas.

Family SILURIDÆ.

Genus AMIURUS.

27. *Amiurus catus*, Linn.   Catfish.
Hab. North America.

Genus SILURUS.

28. *Silurus glanis*, Linn.   Sly Silurus.
Hab. Europe.

Family CYPRINIDÆ.

Genus CYPRINUS.

29. *Cyprinus carpio*, Linn.   Common Carp.
Hab. British fresh waters.

P

30. *Cyprinus gibelio,* Bloch.   Prussian Carp.
    Hab. British fresh waters.

31. *Cyprinus auratus,* Linn.   Gold Carp.
    Hab. British waters (introduced).

Genus COBITIS.

32. *Cobitis barbatula,* Linn.   Common Loach.
    Hab. British fresh waters.

33. *Cobitis fossilis,* Linn.   Thunder-fish.
    Hab. Baltic Sea.

Genus GOBIO.

34. *Gobio fluviatilis,* Cuv. et Val.   Common Gudgeon.
    Hab. British fresh waters.

Genus TINCA.

35. *Tinca vulgaris,* Cuv.   Common Tench.
    Hab. British fresh waters.

Genus LEUCISCUS.

36. *Leuciscus vulgaris,* Flem.   Common Dace.
    Hab. British fresh waters.

37. *Leuciscus rutilus* (Linn.).   Roach.
    Hab. British fresh waters.

38. *Leuciscus phoxinus* (Linn.).   Minnow.
    Hab. British fresh waters.

Family SALMONIDÆ.

Genus SALMO.

39. *Salmo fario,* Linn.   Common Trout.
    Hab. British fresh waters.

40. *Salmo salar,* Linn.   Salmon.
    Hab. British waters.

41. *Salmo lacustris*, auct.   Swiss Lake-Trout.
    Hab. Lakes of Switzerland.

42. *Salmo umbla*, Agassiz.   Swiss Charr.
    Hab. Lakes of Switzerland.

43. *Salmo trutta*, Linn.   Salmon-Trout.
    Hab. British waters.

Family GADIDÆ.

Genus MOTELLA.

44. *Motella mustela* (Linn.).   Five-bearded Rockling.
    Hab. British Seas.

Genus MORRHUA.

45. *Morrhua callarias* (Linn.).   Variable Codfish.
    Hab. Baltic Sea.

Genus LOTA.

46. *Lota vulgaris*, Jenyns.   Burbot.
    Hab. British fresh waters.

Family ESOCIDÆ.

Genus ESOX.

47. *Esox lucius*, Linn.   Pike.
    Hab. British fresh waters.

Family MURÆNIDÆ.

Genus CONGER.

48. *Conger vulgaris*, Cuv.   Conger Eel.
    Hab. British fresh waters.

Genus GYMNOTUS.

49. *Gymnotus electricus*, Lind.   Electric Eel.
    Hab. British Guiana.

Genus ANGUILLA.

50. *Anguilla vulgaris*, Cuv.   Common Eel.
    Hab. British fresh waters.

Family SYNGNATHIDÆ.

Genus SYNGNATHUS.

51. *Syngnathus typhle*, Linn.   Deep-nosed Pipe-fish.
    Hab. British waters.

52. *Syngnathus ophidion*, Linn.   Straight-nosed Pipe-fish.
    Hab. British Seas.

53. *Syngnathus lumbriciformis* (Willughby).   Worm Pipe-
    fish.
    Hab. British Seas.

54. *Syngnathus æquoreus*, Linn.   Æquoreal Pipe-fish.
    Hab. British Seas.

PROCEEDINGS OF THE SCIENTIFIC MEETINGS OF THE ZOOLOGICAL SOCIETY OF LONDON.

8vo.

	Complete.		Letterpress only.		Illustrations only.	
	To Fellows.	To the Public.	To Fellows.	To the Public.	To Fellows.	To the Public.
1861, cloth	32s	47s	4s. 6d	6s	27s. 6d	41s.
1862, ,,	32s	47s	4s. 6d	6s	27s. 6d	41s.
1863, ,,	32s	47s	4s. 6d	6s	27s. 6d	41s.
1864, ,,	32s	47s	4s. 6d	6s	27s. 6d	41s.
1865, ,,	32s	47s	4s. 6d	6s	27s. 6d	41s.
1866, ,,	32s	47s	4s. 6d	6s	27s. 6d	41s.

	With Illustrations, Uncoloured.		With Illustrations, Coloured.	
1867, Part I	3s. 0d. each	4s. each	10s. each	15s. each.

TRANSACTIONS OF THE ZOOLOGICAL SOCIETY OF LONDON. 4to. 5 vols.

		To Fellows.			To the Public.		
		£	s.	d.	£	s.	d.
Vol. I., containing 59 Plates	... ... Price	3	13	6	... 4	18	0
Vol. II., ,, 71 ,,	... ... ,,	4	0	0	... 5	6	6
Vol. III., ,, 63 ,,	... ... ,,	3	8	6	... 4	11	0
Vol. IV., ,, 78 ,,	... ... ,,	6	2	0	... 8	2	6
Vol. V., ,, 67 ,,	... ... ,,	5	3	6	... 6	19	0
Vol. VI., Part I. 14 ,,	... ... ,,	1	7	0	... 1	16	0
Vol. VI., ,, II. 10 ,,	... ... ,,	1	7	0	... 1	16	0
Vol. VI., ,, III., 6 ,,	... .. ,,	1	7	0	... 1	16	0

The following are the most recently published Parts of the "Transactions":—

Vol. VI. Part 1. Price 36s. "On the Characters and Affinities of *Potamogale*," by Prof. G. J. Allman; "On some Indian Cetacea collected by Walter Elliot, Esq.," by Prof. Owen, F.R.S.

Vol. VI. Part 2. Price 36s. "On the Osteology of the Dodo," by Prof. Owen, F.R.S.

Vol. VI. Part 3. Price 36s. "Description of the Skeleton of *Inia geoffrensis* and of the Skull of *Pontoporia blainvillii*," by William Henry Flower, F.R.S.; "On a Raptorial Bird transmitted by Mr. Andersson from Damara Land," by J. H. Gurney, F.Z.S.; "On some Fossil Birds from the Zebbug Cave, Malta," by W. K. Parker, F.R.S.

www.ingramcontent.com/pod-product-compliance
Lightning Source LLC
Chambersburg PA
CBHW021704210326
41599CB00013B/1514